Asphalt Mixture Selection

Asphalt Mixture Selection

Cliff Nicholls and
Arthur Hannah

CRC Press
Taylor & Francis Group
Boca Raton London New York

CRC Press is an imprint of the
Taylor & Francis Group, an **informa** business

CRC Press
Taylor & Francis Group
6000 Broken Sound Parkway NW, Suite 300
Boca Raton, FL 33487-2742

First issued in paperback 2021

ISBN-13: 978-1-138-61682-0 (hbk)
ISBN-13: 978-1-03-217734-2 (pbk)
DOI: 10.1201/9780429461835

Library of Congress Cataloging-in-Publication Data

Names: Nicholls, Cliff, author. | Hannah, Arthur, author.
Title: Asphalt mixture selection / Cliff Nicholls, Arthur Hannah.
Description: First edition. | Boca Raton : CRC, Taylor & Francis, 2019. | Includes bibliographical references and index. | Summary: "This practical guide starts with a survey of the types of site and the asphalt properties which are required. Various external influences which may affect the relative importance of some properties are addressed, and the interplay of sites and external is considered. Asphalt mixture types and their properties are reviewed, largely as defined in the EN 13108 series but subdivided into further categories, and into maximum nominal coarse aggregate sizes using EN 13043 basic set plus set 2 sizes. Guidance is given, including using flowcharts, of the different mixtures that are suitable for each situation. In some cases a range of choices or mixtures with different degrees of suitability is offered. The guidance covers surface course, binder course and base, but with more focus on the surface course where the external influence is most significant. The site and external influence combinations on which a mixture can be used successfully are also given. The book is primarily intended for those who select asphalt on an occasional basis, such as architects or housing developers, but could be of use to other engineers with limited experience. It is also useful as an educational textbook for those studying asphalt technology"– Provided by publisher.
Identifiers: LCCN 2019023318 (print) | LCCN 2019023319 (ebook) |
ISBN 9781138616820 (hardback ; acid-free paper) | ISBN 9780429461835 (ebook) Subjects: LCSH: Asphalt emulsion mixtures.
Classification: LCC TE275 .N53 2019 (print) | LCC TE275 (ebook) |
DDC 625.8/5–dc23
LC record available at https://lccn.loc.gov/2019023318
LC ebook record available at https://lccn.loc.gov/2019023319

Visit the Taylor & Francis Web site at
http://www.taylorandfrancis.com

and the CRC Press Web site at
http://www.crcpress.com

This book is dedicated to our wives, Carol and Clare, despite them not being in the least interested about the nuances of the subject.

Contents

Tables

Figures

Preface

Asphalt is a complex material that has many beneficial structural and serviceability properties for use in the various layers of road, airfield and other pavements. The differing situations where asphalt is used requires different properties in order to be able to perform satisfactorily. However, there are several different types of asphalt which have different properties, and these properties can be adjusted by the design of the mixture. Therefore, the selection of type of asphalt for each layer is important in the design of a pavement. There is not necessarily an optimal mixture type for every situation but for each situation there is a set of mixture types that are more suitable than others.

A brief description is given of the basic aspects of asphalt mixtures (Chapter 2) followed by brief descriptions of the various properties that can be required of asphalt (Chapter 3), which can vary between pavement layers and local situations. The asphalt mixture types are then reviewed (Chapter 4) using the terminology defined in the EN 13108 series of specifications but subdivided into further categories and into nominal maximum aggregate sizes (NMAS) using EN 13043 basic set plus Set 2 sizes. Reviews of the various types of site that are regularly paved with asphalts (Chapter 5) follow, including advice on which properties will be required of the asphalt for that type of site. This review is followed by another on the various external influences that may affect the relative importance of some properties (Chapter 6). It is a combination of the sites and external influences that produce the myriad of different properties required for different situations in which asphalt is used. These five chapters will form the basis for the following guidance.

Guidance is given, including using flow-charts and tables, on the different mixtures that are suitable for each site/external influence combination (Chapter 7). Mixtures with different degrees of suitability are often included as well as multiple choices being offered. The guidance covers surface course, binder course and base, although most advice is on surface course because the external influence mainly affects the top layer. The converse of the site/external influence combinations on which a mixture can be used successfully is also given in the form of lists (Chapter 8). The book finishes with examples of asphalt selection and a final summary (Chapter 9).

The book is primarily intended for those who select asphalt on an occasional basis, such as architects or housing developers, but could be of use to other engineers with limited experience. The selection also does not include the need for using high-friction surfacings or other such overlay treatments.

About the Authors

 **Dr Cliff Nicholls,** known generally as Cliff, was educated at King's School Worcester and Imperial College of Science and Technology. Cliff graduated in Civil Engineering from Imperial College in 1972 to join Rendel Palmer & Tritton, before moving to the Property Services Agency, the Department of the Environment, the Building Research Establishment and finally the Transport and Road Research Laboratory (now TRL Limited). From 1988, he has been researching into pavement materials, in particular asphalt and asphalt test methods. Cliff retired from TRL in October 2015 although he remains on a call-off contract. He also does consultancy work through IRD QSTP-LLC and The Driven Company Associates Limited.

Prior to retirement, Cliff was a Senior Academy Fellow in the Infrastructure Division at TRL. He was mainly involved in research into asphalt surface course materials, often as Project Manager. Projects have involved a wide range of materials including hot rolled asphalt to surface dressing and including porous asphalt, high-friction surfacings, thin surface course systems and stone mastic asphalt as well as some on associated materials such as road markings.

Cliff sat on British Standards Institution and Comité Européen de Normalisation (CEN) committees for asphalt, including being the convenor of the CEN test methods task group for test methods from 2000 to 2015. He was also a member of several British Board of Agrément Highway Authorities Products Approval Scheme steering groups and was a member of the Council of the Institute of Asphalt Technology. Cliff has written many TRL reports and other learned papers. He edited the book *Asphalt Surfacings*, which was published by E & FN Spon in 1998, and has contributed to other books. He obtained a DPhil by published works from the University of Ulster in August 1999.

 **Arthur Hannah** worked for RMC, subsequently CEMEX, and has experience covering numerous senior management positions in the asphalt industry over 45 years, covering all production, commercial, contracting and technical disciplines, working for some of the major players in the industry. He has gained considerable experience in site inspections and the resolution of problems in the asphalt production, surfacing and quarrying industries and in setting up quality management systems and the subsequent auditing and gaining approval. He has advised clients on the use of asphalt solutions and the options available to meet their specific requirements.

Arthur moved to TRL Limited, when he worked on numerous assignments in Africa and the Middle East looking into material design and their application in extreme climatic conditions. He has presented training courses on a variety of asphalt subjects at many levels. Whilst at TRL, he became President of the UK Institute of Asphalt Technology from 2013 to 2015 and was awarded Honorary Fellowship in 2019. Arthur is now an independent consultant as Fifty 2 Consultancy.

Chapter 1

Introduction

1.1 Objective

Asphalt is a pavement material that can provide multiple properties to a greater or lesser extent (Nicholls, 2017). As such, it is used in a variety of situations that include roads, airfields, footways, cycleways, parking areas, docks and other ancillary purposes that present different challenges. The challenges are further complicated by the traffic (weight and frequency), geometry (including hills, bends and intersections) and climate.

However, asphalt is not so much a material as a family of materials, as shown by the number of different types of asphalt covered by EN 13108, the European standard for asphalt mixtures (CEN, 2016), which can be further subdivided as shown in the UK guidance on EN 13108 (BSI, 2016). These variations are further multiplied by the choice of aggregate source and binder type and grade.

Given the wide range of challenges and mixtures, knowledge is needed to select an appropriate mixture for each set of challenges. Such knowledge is built up with experience, but the authors are not aware of any consistent methodology that has been developed to assist those people with limited experience to make the right choice. Therefore, a simple methodology has been developed from reviews of the challenges set by different situations and of the different properties inherent of the different mixtures. This methodology can be used to identify appropriate mixtures for particular situations and the situations for which a particular mixture is suitable. However, the methodology does not necessarily produce a single 'correct' mixture but one or more different mixtures that are appropriate.

In this methodology, the category of aggregate will be included but not the various quarries which supply those categories in order to keep the methodology international. The use of quarries would also make the methodology unmanageable because of the large number of possible quarries in the UK alone.

The inclusion of reclaimed asphalt will not be explicitly covered because this component material should not affect the properties of an asphalt mixture or the situations in which it can be used. The only difference is likely to be the environmental and financial cost of the material. Therefore, each mixture type covered will include both such mixtures with and without reclaimed asphalt.

1.2 Terminology and Units

Oscar Wilde described the UK and USA as 'two nations divided by a common language' and nowhere is this statement more relevant than for asphalt technology. The two countries use different terms to mean the same thing for several items, and these differences have been further exacerbated by the harmonisation of terms across Europe requiring the UK to change some terms that were previously the same as those used in the USA. Therefore, it is important to clarify which terms are being adopted in any publication.

The main differences are with regard to the materials and the layers in which that material is used. The term 'asphalt' is used in Europe solely to mean the mixture including the aggregate whereas in America it can be used to mean either the mixture, when it is generally introduced as 'asphalt concrete', 'asphaltic concrete' or 'hot mix asphalt (HMA)' or the binder, when it is generally introduced as 'asphalt cement'. In Europe, 'asphalt concrete' is a specific mixture type (Sub-Section 4.1.2) which can cause confusion when it is uncertain whether the term relates to the mixture type or to all/any of them, while the binder is known as 'bitumen'.

With regard to layers, the European terminology is 'surface course' at the top, 'binder course' next and 'base' (which may be split into upper base and lower base) whereas the American terms, which were previously used in the UK, are 'wearing course', 'basecourse' and 'roadbase'. There is potential for confusion if the term 'base course' is used as to whether it is the second layer with the space omitted or the bottom layer with 'course' erroneously added.

The terms are listed in Table 1.1 for quick reference. For this book, the European terms will generally be used unless stated otherwise.

The term 'hot mix asphalt' (HMA) has been widely used for referring to asphalt world-wide. However, the term has become more specific with the recent increasing use of asphalt mixtures that are mixed and laid at temperatures lower than those that were previously standard. These new reduced temperature mixtures are generally categorised as 'warm mix asphalt', 'half-warm mix asphalt' (or 'semi-warm mix asphalt') and 'cold mix asphalt'.

With regard to units, this book has been written with all units converted to SI units (which excludes centimetres, despite the unit sometimes being mistakenly used even in standards) whenever possible to keep the units

Table 1.1 Conflicts in Terminology

Description	European		American
Mixture	Asphalt		Asphalt(ic) concrete
Recycled mixture	Reclaimed asphalt (RA)		Reclaimed asphalt pavement (RAP)
Binder	Bitumen		Asphalt cement
Top layer	(Surface course	Surfacing	Wearing course
Second layer	Binder course)		Basecourse
Bottom layer(s)	Base		Roadbase

consistent. Therefore, where American tests are referenced, it is the metric conversion that has been quoted rather than the imperial units. Also, the aggregate sizes used are those in Basic plus Set 2 in accordance with EN 13,043 (CEN, 2002); these sizes are 0, 1 mm, 2 mm, 4 mm, 6.3 (6) mm, 8 mm, 10 mm, 12.5 (12) mm, 14 mm, 16 mm, 20 mm, 31.5 (32), 40 mm and 63 mm but it is intended any selection by aggregate size includes the nearest equivalent size in Basic plus Set 1 in accordance with EN 13043 or other set of aggregate size.

References

British Standard Institution (2016). Guidance on the use of BS EN 13108 "Bituminous mixtures – Material specifications". *PD 6691:2015+A1:2016*. London: British Standard Institution.

Comité Européen de Normalisation (2002). Aggregates for bituminous mixtures and surface treatments for roads, airfields and other trafficked areas. *EN 13043:2002*. Most European standardisation institutions.

Comité Européen de Normalisation (2016). Bituminous mixtures – Material specifications. *EN 13108-1:2016, EN 13108-2:2016, EN 13108-3:2016, EN 13108-4:2016, EN 13108-5:2016, EN 13108-6:2016, EN 13108-7:2016, EN 13108-8:2016, EN 13108-9:2016, EN 13108-20:2016, EN 13108-21:2016*. Most European standardisation institutions.

Nicholls, J C (2017). *Asphalt Mixture Specification and Testing*. London: CRC Press.

Chapter 2

Basics

2.1 Structure of the Pavement

A flexible pavement is built up in layers of bound asphalt over a foundation of unbound aggregate, again often in several layers. The bound layers are termed the base, the binder course and the surface course. In some situations, the surface course is covered with surface treatments, in particular surface dressings, slurry surfacings and high-friction surfacing. The selection of surface treatments is not covered in this book.

The base is the lowest bound layer of a flexible pavement, although it is not always laid in a single lift. The main purpose of the base is to provide the structure to distribute all the loading applied to the pavement down to the unbound foundation. The strength of the foundation is also important in order to be able to receive those distributed loads without damage. The thickness of the base layer is dependent on the loading applied and the strength of the asphalt and can be several hundred millimetres in total.

The binder course is a nominal part of the surfacing but it also contributes to the structural strength of the pavement. Therefore, it has to be designed against a wider range of criteria than other layers. The binder course was traditionally 60 mm thick with a 40 mm surface course to provide 100 mm of surfacing, but alternative surface course types with different thicknesses have been developed so that the binder course is often adjusted to compensate. However, there is no definitive reason for fixing the surfacing thickness to 100 mm provided the lower layers can take the more direct stresses when any distribution of the load is less effective due to less surfacing.

The surface course is the layer in direct contact with tyres of the vehicles or the shoes of people using the pavement. As such, the asphalt needs a whole series of properties in addition to those required of asphalt in the other layers. However, the surface course is not considered to contribute to the structural strength of the pavement because it needs replacing most frequently and the contribution could be different for future replacements. Surface courses were traditionally 40 mm thick, but they were increased for hot rolled asphalt to minimise loss of chippings due to premature cooling during construction and were reduced in other materials to minimise the use of material, particularly scarce high polished stone value aggregate.

The aggregate size of the layers tends to decrease in higher layers because of the need for a more level surface to produce good ride quality. The smaller the aggregate size, the closer control can be achieved.

2.2 Compositional Aspects

2.2.1 Aggregate Type

The aggregate type will affect the properties that the aggregate particles in a mixture will have. Several aggregate properties are important to the performance of the mixture, such as the strength being important for heavily trafficked pavements and the retained micro-texture being important for the surface course layer. However, these properties will not be considered here because the aggregate type can be selected from those locally available that have the appropriate aggregate properties once the mixture type has been selected.

The use of reclaimed asphalt (RA) to replace some of the aggregate should not affect the properties of the mixture type if the mixture is properly designed and the quality control is sufficient. Therefore, the use of RA will not be covered explicitly although its use does generally improve the sustainability.

2.2.2 Nominal Maximum Aggregate Size

The nominal maximum aggregate size (NMAS) of a mixture can be a consideration in the selection of the appropriate mixture type. Therefore, the selection of appropriate mixture types will include its NMAS value when relevant.

The NMAS needs to be restrained as a proportion of the thickness in which the mixture is laid. The need for the limitation is to avoid the mixture dragging during laying with piles of large particles exceeding the nominal layer thickness. The traditional ratio was that the NMAS should be no greater than 40% of that thickness, but more recently introduced mixtures such as asphalt for ultra-thin layers (AUTL) (Section 4.5) are designed for successful laying with the NMAS at a greater proportion of the layer thickness.

The NMAS may also need to be restrained as a proportion of the thickness in which the mixture is laid but as a minimum rather than maximum. Some mixture types will have reduced deformation resistance because the NMAS is very much less than the layer thickness in which it is to be compacted. An alternative to increasing the NMAS is to lay the asphalt in more layers, but this policy increases the number of horizontal joints with the concomitant risks.

Reducing the NMAS generally improves the resistance to scuffing of a mixture. Therefore, mixtures with smaller NMAS are often used for roundabouts and other areas of high stress.

Increasing the NMAS of a mixture will reduce the required binder content because there will be a reduced surface area of the aggregate particles that need to be coated by the bitumen. The reduced binder content will reduce the cost but will have a detrimental effect on the durability, particularly if the mixture is used on sites with water regularly present and/or the aggregate type is particularly hydrophilic.

2.2.3 Aggregate Grading

The aggregate grading can vary from single-sized, in which most of the aggregate falls within a single fraction, to continuous, in which each fraction tries to fill the gaps left between the particles in the larger fractions to produce a denser mixture. Other types are gap-graded, in which some of the fractions are effectively missing in order to get better aggregate interlock, and the use of a large fraction to bulk up a small aggregate mixture without much interlock between the large particles.

2.2.4 Binder Content

The underlying feature is the binder film thickness which is the binder content divided by the surface area of the aggregate. The surface area can be estimated but will be strongly dependent on the particle shape and the extent to which that shape is regular. Therefore, the binder content is generally used rather than the binder film thickness.

The binder film thickness/binder content has an effect on several properties of a mixture, with increased contents increasing the durability and ductility but reducing the mechanical properties. However, the binder content will not be included in the selection process here because it would make the procedure too detailed.

2.2.5 Binder Grade

The grade of binder is often dictated by the type of mixture, as a limited range of grades if not as a single grade. The use of a harder grade will tend to increase the mechanical properties of the mixture but reduce the resistance to fatigue and potential durability. Similarly, the addition of polymer modifiers to the bitumen can be dictated by the mixture type, such as asphalt concrete for very thin layers (BBTM) (Section 4.4), but the addition can also be used to remediate deficiencies in some properties of the mixture. Therefore, the grade and/or type will be included in the type of mixture selected when relevant.

2.2.6 Air Voids Content

The air voids content is an important factor in the performance of an asphalt because it indicates the extent that air and water can permeate the mixture. The air voids content can be considered to be just a combination of the aggregate grading and the binder content but that combination is a useful parameter in its own right. The in-situ air voids content is also a measure of the effectiveness of the compaction when compared to the design air voids content measured in the laboratory.

Chapter 3

Asphalt Properties

3.1 General

This chapter reviews the principal properties required from asphalt. Many of the properties could be classed as resistance to common defects that can develop in asphalt pavements whilst others are distinct properties required in themselves, particularly of the surfacing. The extent to which these properties are important will vary depending on a number of factors, of which the layer within the pavement is a particularly important one. The discussion here is about the properties actually required rather than the specification of an appropriate measure of them, which is covered elsewhere (Nicholls, 2017).

The properties are grouped into properties generally applicable to all layers, those applicable to the structural layers, those applicable to the surface course and those applicable to the pavement generally.

3.2 Common Properties

3.2.1 Resistance to Deformation

Structural deformation is the result of the failure of the foundation as a result of the pavement not being able to distribute the applied wheel-loads sufficiently widely and the stresses imposed at the top of the foundation being greater than that which the foundation could withstand. This form of deformation results in a wide deformation across practically the full width of the lane. The inherent property required to avoid such deformation is adequate strength of the pavement which, in turn, is the product of the asphalt stiffness (Sub-Section 3.3.1) and the square of the pavement thickness provided each layer of the pavement is fully bonded to its adjacent ones. If the layers are not bonded, the strength of the pavement is the sum of the strength of each layer calculated separately. As such, resistance to structural deformation is part of the design of the pavement rather than of the asphalt mixture.

Surface deformation, as shown in Figure 3.1, is a more local form of deformation that is located along the wheel-tracks and resistance to it is part of the design of the asphalt mixture. The deformation occurs as the material under the wheel path is further compacted and/or is pushed to either side. It usually affects approximately the top 100 mm of the pavement, generally the surface and binder courses, with more resistance being required the nearer the surface. The resistance to this type of deformation is a combination of the interlock of the coarse aggregate particles and the stiffness of the mortar (a combination of the binder, filler and fine aggregate), with the contribution of each depending on the type of mixture. The aggregate interlock will depend on the shape of aggregate particles as well as proportion of the mixture that is coarse aggregate. The contribution of mortar is very temperature-dependent so that any deformation tends to occur just on the hottest few days of the year, particularly for mortar-based mixtures.

The typical influences that the compositional aspects have upon the resistance to deformation of an asphalt mixture are:

- Aggregate type – Crushing value is important for strength.
- Maximum nominal aggregate size – Larger size will resist deformation better.
- Aggregate grading – Interlocking of aggregate is crucial.
- Binder content – Higher binder contents aid lubrication so aid rutting.
- Binder grade – Softer binders will deform more easily so aid rutting.
- Air voids content – More air voids allow for more secondary compaction and, therefore, aid rutting.

3.2.2 Resistance to Cracking

Traditionally, cracking was assumed generally to start at the bottom of the pavement and travel upwards because, considering the pavement as

Figure 3.1 Cross-section through a rut.

a beam that is loaded by being bent downwards at the centre, the highest stresses would be at the bottom. However, it has been found that cracks in thick pavements generally travel from the top downwards (Leech and Nunn, 1997). The advantage of cracks being top-down rather than bottom-up is that, if the crack does not propagate down too quickly, the pavement can be repaired by just replacing the depth to which the crack has developed.

There are several different forms of cracking, including:

Fatigue (alligator) cracking: The cracks (Figure 3.2) are caused by repeated recoverable deflections of the asphalt layer(s) that weaken the structure with time. The amplitude of the deflections is dependent on the traffic loads imposed and the strength of the pavement layers, particularly the asphalt layers for which the strength is proportional to the material stiffness and the depth. The strength of the asphalt layers can be based on the combined depth if there is adequate bond but needs to be calculated for each layer separately and then summed if there is no bond: in practice, it is usually somewhere between the two extremes. The cracking emerges as a pattern of irregular cracks that criss-cross the surface and form irregular blocks that look like alligator scales; hence the alternative name of alligator cracking (Nikolaides, 2015).

Thermal fatigue cracking: The cracks develop from thermal cycling when the binder becomes too stiff to withstand the thermally induced

Figure 3.2 Example of fatigue (alligator) cracking.

stress and, as such, is related to the thermal expansion coefficient and the relaxation characteristics of the mixture together with the age of the pavement as the binder hardens as a result of oxidation or time-dependent physical hardening. The cracks take several seasons to propagate through the asphalt layers, initiated at the surface and propagating relatively slowly with each thermal cycle (Whiteoak and Read, 2003).

Linear wheel-track cracking: This cracking is a form of fatigue and has the same causes. However, it develops along the wheel tracks of the near-side lane, not always simultaneously, and consists of a single longitudinal crack with small branches (Nikolaides, 2015).

Reflective cracking: The cracks (Figure 3.3) occur in overlays to pavements that have joints or significant existing joints (typically rigid pavements, flexible pavements with cement-bound bases, tied shoulders and widenings). They result from relative vertical and/or horizontal movements either side of the underlying discontinuities caused by different responses to traffic loading, temperature compression/expansion, moisture changes and/or swelling/shrinkage of the subgrade. The pattern of cracks tends to follow the pattern of the underlying discontinuities, as shown in Figure 3.3 (Nikolaides, 2015).

Paving joint and widening cracking: The cracks appear at construction joints between paving lanes or at the joints produced when pavement widening. Initially, the cracks are longitudinal along the direction of the joint but small branch cracks may develop later. The cause can be lack of material at the joint, lack of bond between different materials and/or poor compaction. Temperature differences between materials during paving and thermal stresses during the service life of the pavement can also be contributory factors (Nikolaides, 2015).

Edge cracking: The cracks are longitudinal and appear at a distance of about 0.3 m to 0.5 m from the pavement edge with or without branching cracks towards the shoulder. The cracking is the result of poor compaction, shrinkage, poor drainage and/or frost action causing there to be insufficient pavement support (Nikolaides, 2015).

Figure 3.3 Examples of reflective cracks.

Low-temperature cracking: The cracks result when the binder becomes too stiff because of extreme cold, generally below about −30°C. The cracks are transverse, run the full depth of the pavement and can appear very quickly (Whiteoak and Read, 2003).

Slippage cracking: The cracks are formed by slippage of the asphalt over the substrate from lack of bond between the layers allowing delamination. The cracks are in the shape of a crescent or half-moon (Nikolaides, 2015).

Helical or diagonal cracking: The cracks are caused by instability or settlement of embankments. The cracks usually start from the centre of the pavement and run towards the pavement edge, diverge from the longitudinal axis or run diagonally across the pavement (Nikolaides, 2015).

The resistance of asphalt to cracking involves two aspects, the resistance to crack initiation and the resistance to crack propagation. These two properties are not necessarily related, with some mixtures developing cracks relatively quickly but those cracks travelling down slowly while other mixtures take some time before any cracks are initiated but then the cracks become full depth almost immediately.

The typical influences that the compositional aspects have upon the resistance to cracking of an asphalt mixture are:

- Aggregate type – Has little effect on cracking.
- Maximum nominal aggregate size – Has limited effect on cracking, but larger nominal size will be more prone to cracking.
- Aggregate grading – Fine-graded material will resist rutting better than a coarse grading.
- Binder content – Low binder contents make stiffer mixtures that are more prone to cracking.
- Binder grade – Softer grades will resist cracking better than harder grades.
- Air voids content – Fewer air voids will resist cracking better than more air voids.

3.2.3 Permeability

Water (Sub-Section 3.2.4) and oxidation (Sub-Section 3.5.1) can affect various properties of the asphalt. Traditionally, the approach to minimising the extent that water and oxygen can have access to the bitumen and the bitumen/aggregate interface of the asphalt is to make the asphalt as impermeable as possible. However, more recently the development of sustainable drainage systems (SUDS) for paved areas, generally with relatively light traffic levels, as well as the need for reduced noise (Sub-Section 3.4.3) and spray (Sub-Section 3.4.4) generation has required asphalt that is as permeable as possible so that the water can pass through as quickly as possible. The voids in such open mixtures are still exposed to greater oxidation than impermeable mixtures. Therefore, the ideal for an asphalt can be either totally impermeable or very permeable depending on the intended use.

The permeability of an asphalt mixture is generally equated to its air voids content. There are two aspects to the air voids content of an asphalt mixtures, the potential air voids content that the mixture can attain and the actual air voids content that the mixture has attained on site. The difference between the two depends on how effective the compaction was, with even permeability needing the difference to be as small as possible for good durability. However, the compaction is a workmanship issue that does not affect the selection process of the mixture type other than in terms of the ease of compaction.

The typical influences that the compositional aspects have upon the permeability of an asphalt mixture are:

- Aggregate type – Has little effect on permeability.
- Maximum nominal aggregate size – Higher nominal size will increase the permeability.
- Aggregate grading – A more single-sized grading will increase permeability.
- Binder content – The binder content needs to be carefully controlled in order to prevent drainage from the binder film around aggregate into the air voids.
- Binder grade – Has little effect on the permeability in service but a softer grade will aid the laying of the material.
- Air voids content – Many air voids are essential for permeability.

3.2.4 Resistance to Water Damage

To varying degrees, most aggregate particles prefer to have water rather than bitumen on their surface. If the internal bond between the binder and aggregate particles weakens in the presence of water, the asphalt mixtures is considered to be susceptible to moisture. The aggregates with a high silica content tend to be more hydrophilic with the order of typical aggregate types being quartzite, granite, sandstone, dolerite, basalt, limestone and marble, the most hydrophobic of them (Hunter et al., 2015). The resistance to water damage can be enhanced by additives.

If water displaces the bitumen on the face of the aggregate particles, it will cause 'stripping' and the breakdown of cohesion of the mixture (Sub-Section 3.4.2). The extent of damage will depend on the ease of access of water to the aggregate particle surface and the extent to which the aggregate is hydrophilic. The ease of access will in turn depend on the continuity of the binder film around each particle and, other than at the surface, the permeability of the mixture (Sub-Section 3.2.3). The damage caused by moisture damage, even if not causing the overall dis-integration, will generally diminish most of the properties of the asphalt.

Resistance to water damage is included in the common properties because water can access the pavement from above (including from precipitation), below (from the water table) or the side (including from inadequate drainage systems).

The typical influences that the compositional aspects have upon the resistance to water damage of an asphalt mixture are:

- Aggregate type – Aggregate type is critical because the aggregate needs to have a strong affinity for bitumen to ensure a good bond between the two.
- Maximum nominal aggregate size – Aggregate size has limited effect on resistance to water damage.
- Aggregate grading – A dense grading will help the asphalt mixture resist water damage.
- Binder content – A high binder content, giving a thicker binder film on the aggregate, will resist water damage better than a low binder content material.
- Binder grade – A harder grade will reach the critical point of damage quicker due to oxidation and, therefore, will not protect the aggregate from water ingress for as long in service.
- Air voids content – A high air voids content will make the material more susceptible to water damage.

3.3 Structural Properties

3.3.1 Stiffness

Stiffness is the extent to which a material will resist deformation in response to an applied force and is the main parameter in pavement design because it is related to the load-spreading ability of the structure, with the base and binder course being the main load-spreading layers of the road. Load spreading reduces the stresses and strains developed in the sub-base and subgrade by traffic loading and influences the level of tensile stress at the underside of the base, which is considered to be an indicator of the risk of fatigue cracking (Nunn and Smith, 1997).

However, the stiffness of asphalt is not as simple a property as it is for most structural materials because asphalt is visco-elastic in that it responds to loading in both a viscous and an elastic manner (Hunter et al., 2015). The extent that the response is viscous or elastic depends on the loading time and the temperature. The binder tends to provide the viscous aspect of the stiffness while the aggregate skeleton tends to provide the elastic and plastic aspects.

The typical influences that the compositional aspects have upon the stiffness of an asphalt mixture are:

- Aggregate type – The aggregate needs to have a high crushing strength in order to avoid any damage due to loading.
- Maximum nominal aggregate size – Aggregate size has little effect on the stiffness of the material.
- Aggregate grading – Aggregate grading is critical because good contact between the aggregate particles is necessary for good stiffness.
- Binder content – A lower binder content, within reasonable limits, will give a greater stiffness in the material.
- Binder grade – Binder grade is critical as a softer binder will give a lower stiffness value.
- Air voids content – More air voids will give a lower stiffness value.

3.3.2 Resistance to Fatigue

Fatigue occurs when the asphalt cannot resist the cumulative effect of repeated loads and generally results in the cracking and even disintegration of the layer being loaded and any higher layers. Fatigue resistance is generally achieved by the material being more flexible so that it can sustain repeated deformations, but stiffer materials will deform less under the same loads and, therefore, be damaged to a lesser extent. Unfortunately, the two aspects are effectively mutually exclusive, with resistance to fatigue generally being considered as the ability to absorb repeated deformations rather than minimising their size (Nicholls, 2017). The lack of fatigue resistance results in cracking (Sub-Section 3.2.2).

The resistance to fatigue, alongside stiffness modulus, has long been a primary consideration in mixture design. However, the observation that all cracks in thick pavements travel from the top down rather than from the bottom up (Leech and Nunn, 1997) led to the concept behind long-life pavements, in which cracks are repaired by replacement of just the surface course, which means that the importance of fatigue resistance is less critical for such pavements.

The typical influences that the compositional aspects have upon the resistance to fatigue of an asphalt mixture are:

- Aggregate type – aggregate type has little effect on the fatigue life.
- Maximum nominal aggregate size – Aggregate size has little effect on the fatigue life.
- Aggregate grading – Aggregate grading has a limited effect on fatigue life.
- Binder content – Higher binder content will increase the fatigue life.

- Binder grade – Softer binder will increase the fatigue life.
- Air voids content – Lower air voids content will increase the fatigue life.

3.3.3 Tensile Strength

Tensile strength is a fundamental engineering property of asphalt that is dependent on the temperature and loading rate. However, asphalt is stronger in compression than tension and the stiffness has traditionally been the property used in pavement design. However, the indirect tensile strength is often used as a surrogate for crack resistance because it is the tensile strength that will resist the initiation and propagation of cracks (Sub-Section 3.2.2) and the ratio of the strength before and after exposure to water as a surrogate for the durability (Sub-Section 3.5.2) of a compacted asphalt mixture.

The typical influences that the compositional aspects have upon the tensile strength of an asphalt mixture are:

- Aggregate type – Aggregate type has no effect on tensile strength.
- Maximum nominal aggregate size – Has no effect on tensile strength.
- Aggregate grading – Aggregate grading has little effect on tensile strength.
- Binder content – Both higher and lower binder contents will reduce tensile strength.
- Binder grade – A modified binder will improve tensile strength.
- Air voids content – Both high and low air voids content will reduce tensile strength.

3.4 Surface Properties

3.4.1 Skid Resistance

Skid resistance is a property that is only applicable to surface course materials and, to a lesser extent, to layers that are to be used as temporary surface courses. Skid resistance refers to the contribution that the road surface provides to tyre/road friction, which relates to the friction available to a particular driver, in particular circumstances, at a particular time. It is influenced by both the vehicle and the road surface as well as external parameters such as the weather and localised contamination. The skid resistance usually refers to the skid resistance on a wetted road surface. The main aspects that influence the skid resistance are generally considered to be macro- and micro-texture (Figure 3.4).

Macro-texture, or texture depth, results from the coarser spaces between particles or grooves in the surface of the pavement. Texture

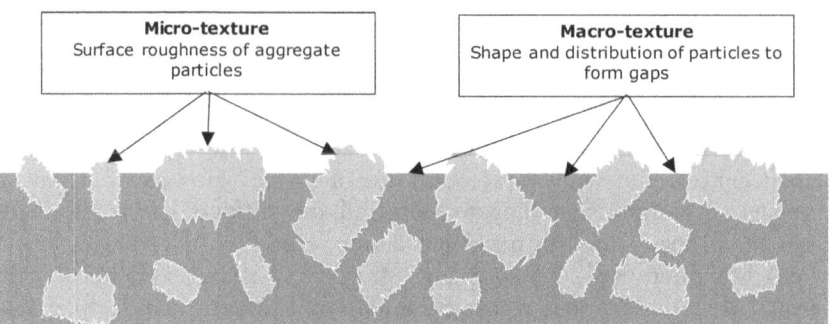

Figure 3.4 Schematic of micro- and micro-texture.

depth contributes to the water drainage, which is important to allow any water to be able to drain away from the tyre/road interface and enable the tyre to grip the micro-texture, which is particularly important if high-speed skidding resistance is to be maintained. Research has also shown both that the accident risk increases with reduced texture depth, even in dry conditions (Roe et al., 1991), and that the texture depth has a strong influence on the relationship between skidding resistance and speed, particularly at low speeds (Roe et al., 1998).

The micro-texture is the fine texture of aggregate particles at the pavement surface that break through the water film in wet conditions and, hence, allows frictional forces to be generated. Micro-texture is the key component in road surface friction without which the road would have hardly any friction in the wet. However, micro-texture can be polished away by traffic, particularly heavy traffic, so that the aggregate type needs to have both the micro-texture and the ability to resist wear. Research (Roe and Hartshorne, 1998) has led to changes in the UK requirements for the polishing resistance of aggregates in new surfaces.

The typical influences that the compositional aspects have upon the skid resistance of an asphalt mixture are:

- Aggregate type – The aggregate texture and rate at which it will wear are important factors.
- Maximum nominal aggregate size – A larger nominal size tends to improve skid resistance.
- Aggregate grading – Aggregate grading is important in obtaining a textured road surface.
- Binder content – High binder contents will reduce skid resistance.

- Binder grade – Binder grade has little effect on skid resistance.
- Air voids content – Air voids are important in terms of surface texture.

3.4.2 Cohesion

Any lack of cohesion of an asphalt mixture can result in the loss of aggregate particles (ravelling) and/or the loss of mortar (fretting) from the surface of the material. The two defects are not independent because the loss of aggregate particles will expose the mortar more directly to the imposed loads as well as the environmental effects, both of which will encourage the loss of mortar, while loss of mortar will reduce the support to aggregate particles and encourage their loss.

There are two types of ravelling (early stage ravelling and ravelling at the end of the service life) with different causes. Early stage ravelling (or fretting) results from inherently low cohesion in the as-laid material due either to poor compaction or to inherently weak bonds between the aggregate particles and the binder. Ravelling that develops at the end of the life-span is particularly prevalent in stone–skeleton mixtures like porous asphalt (Section 4.6) and stone mastic asphalt (Section 4.3) and occurs in those areas that bear the loads from vehicle tyres, not only at crossings or curves but also in straight sections of roads. The main factors that affect end-of-life ravelling are the type of mixture, the air voids content of the asphalt mixture, the fatigue resistance of the mastic part of the mixture (which is influenced by the amount and type of bitumen) and the shape of the bituminous bridges between the stones.

Extensive localised ravelling and/or fretting results in potholes (Figure 3.5), a defect that occurs extensively on all types of road. If the potholes become too extensive on pavement, the defect should be classed as overall disintegration rather than each pothole being an individual defect. The overall disintegration is a form of end-of-life ravelling resulting from (premature) ageing of the binder.

Ravelling is a particular problem for airfields with jet powered aircraft because of the danger of foreign object damage (FOD) to the jet engines when loose particles are ingested. FOD can also occur in other situations, such as taking chips out of windscreens and paintwork, but the damage is generally less catastrophic.

The typical influences that the compositional aspects have upon the cohesion of an asphalt mixture are:

- Aggregate type – The aggregate needs to have an affinity for binder.
- Maximum nominal aggregate size – Aggregate size has a limited effect on cohesion.
- Aggregate grading – Aggregate grading has little effect on cohesion.
- Binder content – The binder content needs to be neither too low nor too high.

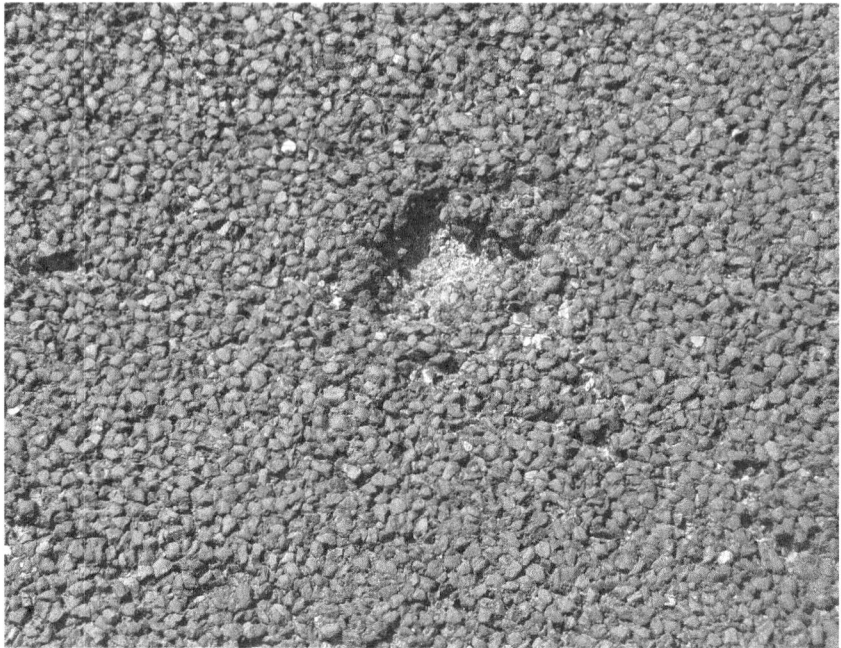

Figure 3.5 Example of a pothole.

- Binder grade – Harder grades will be less cohesive.
- Air voids content – Air voids have little effect on cohesion.

3.4.3 Tyre/Road Noise

Noise resistance is a property that is primarily applicable to surface course materials. Road traffic noise is perceived as a problem by the public, whether travelling on, living near to or otherwise in the vicinity of roads. This perception has increased as vehicles have become quieter and expectations have increased. The properties of the pavement surface of texture depth, flow resistivity and acoustic absorption affect the generation and propagation of tyre/road noise. The other aspect of road noise is the engine noise, with the relative proportion of the two sources varying depending on the composition of the traffic and weather conditions as well as the properties of the pavement (Nicholls, 2017).

The use of open, high texture depth mixtures and small nominal maximum aggregate size are generally considered to be the aspects of the asphalt mixture that can most easily affect the tyre/road noise generated.

The typical influences that the compositional aspects have upon the tyre/road noise of an asphalt mixture are:

- Aggregate type – Aggregate type has little effect on tyre/road noise.
- Maximum nominal aggregate size – Larger aggregate size will generate more noise.
- Aggregate grading – Aggregate grading has an effect on noise due to the texture of the surface.
- Binder content – Binder content has little effect on noise.
- Binder grade – Binder grade has little effect on noise.
- Air voids content – Air voids are important in the that they can dampen noise.

3.4.4 Spray Generation

The extent to which a road surface will generate spray will depend on a number of factors including the intensity of the rainfall, the antecedent rain, the road surface 'shape', the tyre pattern, the load on the tyre and the vehicle speed. The spray generation can change during a rainstorm as the voids in surface shape go from being reservoirs to 'hide' the falling water to 'lakes' from which falling rain will bounce off. Nevertheless, more permeable mixtures are generally considered to generate less spray than less permeable mixtures although there are no standardised test methods to quantify spray generation.

The typical influences that the compositional aspects have upon the spray generation of an asphalt mixture are:

- Aggregate type – Aggregate type has little effect on spray generation.
- Maximum nominal aggregate size – Larger aggregate size will tend to produce more voids to reduce surface water.
- Aggregate grading – Aggregate grading is important in producing voids to reduce surface water.
- Binder content – Binder content has little effect on spray generation.
- Binder grade – Binder content has little effect on spray generation.
- Air voids content – Air voids are important in reducing surface water.

3.4.5 Resistance to Chemical Damage

The damage that can occur from the presence of chemicals tends to be similar to, but generally more severe than, that from the presence of water. However, the presence of chemicals tends to be less prevalent than that of water in most locations. The chemicals that can be found on certain pavements include fuels, lubricating oils, de-icing fluids and animal slurry. Fuels and lubricating oils are found where vehicles are refuelled and serviced and can easily modify the bitumen because they are other hydro-carbon fractions of crude oil. De-icing fluids are widely used on airfields in the winter rather than brine while animal slurries are found on farm roads.

The typical influences that the compositional aspects have upon the resistance to deformation of an asphalt mixture are:

- Aggregate type – Aggregate type has little effect.
- Maximum nominal aggregate size – Aggregate size has little effect.
- Aggregate grading – A denser grading will resist chemical damage more.
- Binder content – A higher binder content will reduce chemical damage.
- Binder grade – A modified binder can help reduce chemical damage.
- Air voids content – Fewer air voids will reduce chemical damage.

3.5 Overall Properties

3.5.1 Adhesion

The adhesion between the aggregate particles and the binder is part of the cohesion of the asphalt, covered in Sub-Section 3.4.2. The adhesion referred to here is the cohesion between layers. As such, the property is a composite property of those of the overlying asphalt, of the substrate asphalt and of any tack or bond coat together with the workmanship during application (in particular the cleanliness).

The structural importance of adhesion is to ensure that all the pavement layers act together as a single 'beam' rather as a series of separate 'beams'. If the layers are not fully bonded, the stiffness (Sub-Section 3.3.1) of the asphalt in each layer needs to be greater or the layer thickness increased to provide adequate load spreading of the loads. A lack of adhesion between layers can also allow water to enter the pavement, permitting that water to damage the asphalt (Sub-Section 3.2.4) with the pressure applied by traffic pumping that water into parts of the joint still bonded.

The adhesion between the binder and surface courses is also important to avoid delamination between the layers with the surface course asphalt coming away in plates (Figure 3.6). The use of thinner surface course layers increases the need for adhesion because there is less mass to help hold the surface course in place.

The typical influences that the compositional have upon the adhesion of an asphalt mixture are:

- Aggregate type – The aggregate needs to have an affinity for binder.
- Maximum nominal aggregate size – Aggregate size has a limited effect on adhesion.
- Aggregate grading – Aggregate grading has little effect on adhesion.
- Binder content – Binder content needs to be neither too low nor too high.

Figure 3.6 Example of delamination.

- Binder grade – Harder grades will be less adhesive.
- Air voids content – Air voids have little effect on adhesion.

3.5.2 Durability

Durability is a property that everybody feels they know about but few can define. However, separate definitions have been proposed for asphalt durability and for pavement durability.

Asphalt durability is defined as (Nicholls et al., 2008):

the maintenance of the structural integrity of compacted material over its expected service life when exposed to the effects of the environment (water, oxygen, sunlight) and traffic loading

Asphalt durability is dependent on:

- the component materials used;
- the weather conditions during laying;
- the mixture, both the generic type and the job mix design;
- the workmanship during mixing, transport, laying and compaction; and
- the site conditions, including geometry, local weather conditions immediately after construction, drainage and (possibly) traffic.

Pavement durability is defined as (Nicholls et al., 2008):

the retention of a satisfactory level of performance over the structure's expected service life without major maintenance for all properties that are required for the particular road situation in addition to asphalt durability

Pavement durability is dependent on:

- the asphalt durability;
- the traffic and other site conditions;
- the performance requirements set; and
- the asphalt performance characteristics.

The durability of the pavement, and hence also of the asphalt, is a critical aspect for the overall performance of the pavement. The importance of durability has increased with the need to improve sustainability. Constructing pavements that do not need to be maintained, or with an increased time between maintenance operations, is often the most sustainable option provided that it does not require significant increases in the use of virgin materials, the consumption of additional energy and/or the carbon footprint (Nicholls et al., 2010).

However, the major problem in estimating the typical service life of asphalt surfacings, if not all layers, is defining the criterion that describes when a surfacing is deemed to be no longer in an acceptable condition for continued use. Thin surfacing sites in the UK monitored for the Highways Agency included some that had remained in service long after an Inspection Panel deemed them unserviceable, whilst others were removed when still in a relatively good condition (Nicholls et al., 2007). The latter included a surfacing that was replaced because it was located within a section for which a major maintenance scheme was being implemented and three surfacings (out of a trial with four different surfacings) that were surface dressed because they were still in a good enough condition to remain serviceable for longer with that dressing (the fourth site, not surface dressed, was in worse condition). Therefore, when numerical criteria such as texture depth and skid resistance are not infringed, the condition when a surfacing would be deemed to be at the end of its service life is not universal. This situation also applies to structural layers, although there are different numeric criteria.

The typical influences that the compositional aspects have upon the durability of an asphalt mixture are:

- Aggregate type – The aggregate needs to have an affinity for binder,
- Maximum nominal aggregate size – Aggregate size has a limited effect on durability.
- Aggregate grading – Aggregate grading has little effect on durability.
- Binder content – Higher binder content will give better durability.
- Binder grade – Binder grade has little effect on durability.
- Air voids content – A denser material will give better durability.

3.5.3 Cost Effectiveness

The price of different asphalts does vary to a certain extent, particularly depending on the cost of the component materials. The aggregates comprise the majority of any mixture with the difference between stone from different sources not being very great. The binder is considerably more expensive than the aggregate so that a higher binder content generally makes the asphalt more expensive. Polymer modifiers will further increase the cost of the mixture, if used, because most are more expensive than the binders despite only consisting of a small proportion of any mixture.

A second aspect that can affect the price of asphalt is the number of properties that the mixture has been designed to achieve and the amount of material over which the cost of that design has to be spread. The quality control required to ensure that those properties are achieved cam also increase the cost of the asphalt as placed.

Another aspect to the cost of the asphalt is the thickness at which it is laid. Improvement to some properties, particularly the stiffness, can allow the thickness to be reduced.

The cost effectiveness of a mixture is the extent to which the additional cost of the asphalt above that of a basic asphalt is warranted by the performance, in particular to the durability if whole-life cost is considered.

The typical influences that the compositional aspects have upon the cost of an asphalt mixture are:

- Aggregate type – Aggregate type has an effect, with higher polished stone value material costing substantially more than limestone.
- Maximum nominal aggregate size – Aggregate size has an effect in that fractions around 10 mm are more sought after and, therefore, command a higher price than other sizes.
- Aggregate grading – Aggregate grading has little impact on cost.
- Binder content – A higher binder content increases the cost.
- Binder grade – Modified binders are more expensive.
- Air voids content – Air voids content has little effect on cost.

References

Hunter, R N, A Self, and J Read (2015). *The Shell Bitumen Handbook* (6th edition). London: Thomas Telford.

Leech, D, and M E Nunn (1997). *Deterioration Mechanisms in Flexible Pavements. Second European Conference on the Durability and Performance of Bituminous Materials*. Leeds: University of Leeds.

Nicholls, J C (2017). *Asphalt Mixture Specification and Testing*. London: CRC Press.

Nicholls, J C, I Carswell, C Thomas, and L K Walter (2007). Durability of thin asphalt surfacing systems: Part 3, findings after six years' monitoring. *TRL Report TRL660*. Wokingham: Transport Research Laboratory.

Nicholls, J C, I Carswell, M Wayman, and J M Reid (2010). Increasing the environmental sustainability of asphalt. *TRL Insight Report INS007*. Wokingham: TRL Limited.

Nicholls, J C, M J McHale, and R D Griffiths (2008). Best practice guide for durability of asphalt pavements. TRL Road Note RN42. Crowthorne: Transport Research Laboratory.

Nikolaides, A (2015). *Highway Engineering: Pavements, Materials and Control of Quality*. London: CRC Press.

Nunn, M E, and T Smith (1997). Road trials of high modulus base for heavily trafficked roads. *TRL Report TRL231*. Wokingham: TRL Limited.

Roe, P G, and S A Hartshorne (1998). The polished stone value of aggregates and in-service skidding resistance. *TRL Report TRL322*. Wokingham: TRL Limited.

Roe, P G, A R Parry, and H E Viner (1998). High- and low-speed skidding resistance: The influence of texture depth. *TRL Report TRL367*. Wokingham: TRL Limited.

Roe, P G, D C Webster, and G West (1991). The relation between the surface texture of roads and accidents. *TRL Research Report RR296*. Wokingham: TRL Limited.

Whiteoak, C D, and J Read (2003). *The Shell Bitumen Handbook* (5th edition). London: Thomas Telford.

Chapter 4

Asphalt Mixture Types

4.1 General

4.1.1 Classification

There are several terms to classify different generic asphalt mixture types, some of which use different terminology for essentially the same mixture type. For consistency, the classification and terminology set out in the EN 13108 series of European standards has been used here. Some categories, in particular asphalt concrete (AC), cover a wide range of different mixtures with the various sub-types of asphalt called by the terminology generally used in the UK for this book (although many of those sub-types have been introduced from elsewhere in Europe).

The main difference between mixture types and sub-types tends to be the aggregate gradings. Other differences are the grade(s) of binder used and the properties required of them (which, to a certain extent, is a circular argument because the properties required are those properties that can be attained with that grading and binder grade). However, the gradings for the different asphalt types are not discrete so that a grading will indicate the mixture type or types that could be intended but gradings do overlap so that it will just confirm which mixture types it is not. As an example, the shaded grading shown in Figure 4.1 could be any of four mixture types with a nominal maximum aggregate size (NMAS) of 10 mm.

In the UK, the term 'thin surface course system' (TSCS) is used widely and many assume that it is the name of a type of asphalt mixture. However, it just indicates that the proprietary mixture has been independently certified for certain properties, despite the mixtures still needing a CE mark which may include many of those properties. Therefore, TSCS will not be included as a separate generic mixture type here.

4.1.2 Mixing Temperature

The full definition of an asphalt mixture now needs to include whether it is 'hot mix asphalt' (HMA), 'warm mix asphalt', 'half-warm mix asphalt'

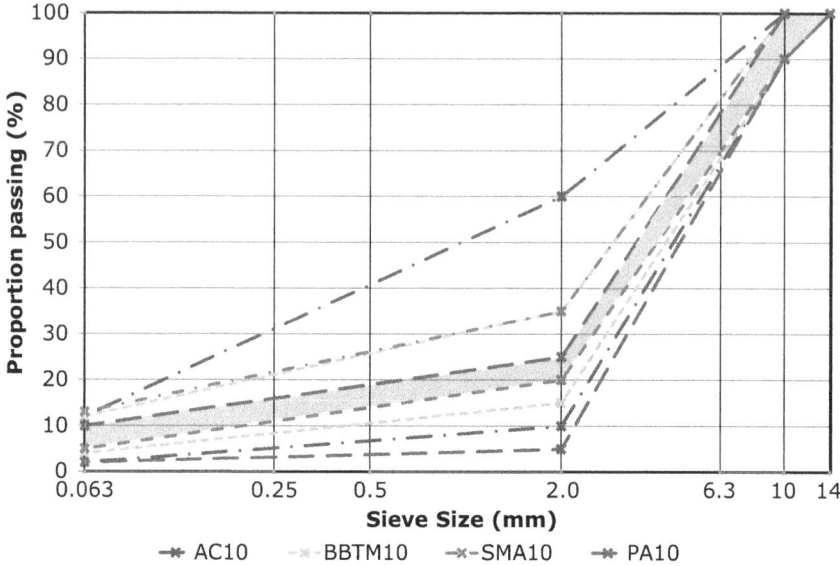

Figure 4.1 Example of overlapping aggregate grading envelopes.

or 'cold mix asphalt'. The distinctions are that HMA is generally mixed at a temperature above 140°C, warm mix asphalt is generally mixed at between 140°C and 100°C, half-warm mix asphalt is generally mixed between 100°C and 70°C and cold mix asphalt is mixed around ambient temperatures, as shown in Figure 4.2.

The lower mixing temperatures can be achieved by several techniques including:

- Organic additives;
- Chemical additives;
- Emulsion-based processes;
- Water-bearing additives; and
- Water-based processes.

However, the lower temperature mixtures are only successful if they replicate the long-term performance of HMA. The one exception is sustainability, where the use of lower temperature mixtures does have the potential for reducing the carbon footprint and energy requirement provided the modification technique in itself does not counterbalance the savings from lower temperature mixing. Given the equivalence, the distinction between the different categories will not be included in the procedure for the selection of appropriate mixtures in this book.

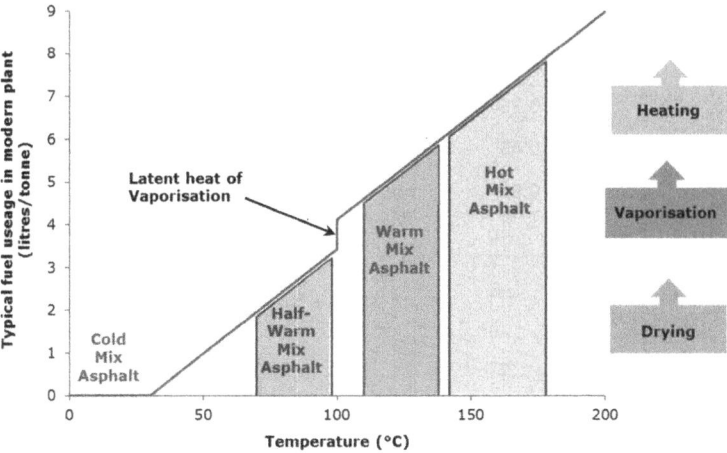

Figure 4.2 Categories of reduced temperature asphalt.

4.2 Asphalt Concrete

4.2.1 General

AC is an asphalt mixture that is standardised in EN 13108-1 (CEN, 2016a). It has the widest grading envelope of any mixture type and, therefore, incorporates many different sub-types with different properties.

The particle distribution of an AC mixture, whether explicitly by design or implicitly from following a standard envelope, is arranged so that the voids between the largest particles are filled with the next smaller size, and so on down through the particle sizes. As such, the particle size distribution roughly follows the Fuller curve as defined in Equation 4.1 and shown in Figure 4.3 (Fuller and Thompson, 1907).

$$P = 100 \times (d/D)^n\% \tag{4.1}$$

where d is sieve size, D is the NMAS of the mixture (in the same units as d), n is the Fuller curve index and P is the proportion of the aggregate passing the sieve size.

The effectiveness of compaction will depend on the shape of the aggregate particles, with a Fuller curve index of 0.5 being the theorical value for round particles but 0.45 generally being found to be more appropriate for 'real' particles. Different AC mixtures have different Fuller curve indices (from 0.35 to 0.6) which produce different AC materials.

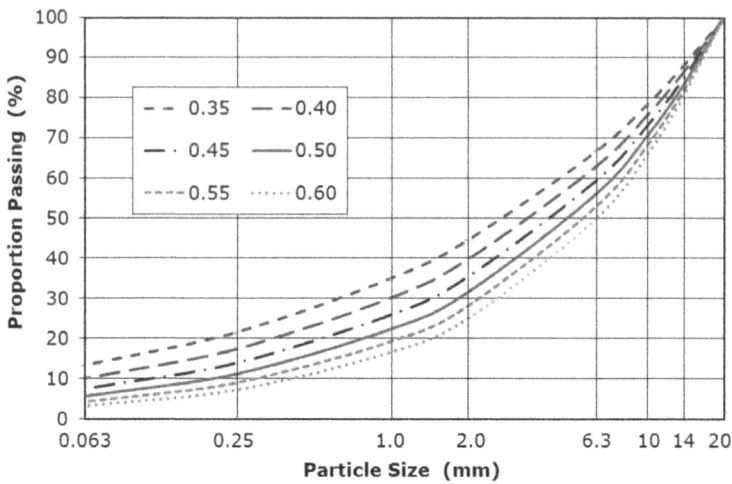

Figure 4.3 Fuller curves for different Fuller Curve indices.

4.2.2 Marshall Asphalt (AC/MA)

AC/MA[1] is a form of AC for which the binder content is optimised using the Marshall method (hence its name), including the use of the Marshall stability test. This test is only permitted under EN 13108-1 for use on airfields and the material is used primarily for airfields, at least in the UK, although it was also used for other applications elsewhere in Europe and the rest of the world. An example specification for military airfields is the UK Defence Estates SPEC 13 (DE MOT, 2009).

The properties used for the design method include Marshall stability, Marshall flow, Marshall quotient, voids in mixture, voids in mineral aggregate and voids filled with bitumen. From these properties, the optimised bitumen content is determined. This content is generally between 4% and 7%, which is higher than the bitumen content generally used for recipe ACs (Sub-Section 4.2.4), making the mixtures denser and, hence, more durable. This enhanced durability is further encouraged by the better quality-control possible with the associated practice of using a mobile asphalt plant dedicated to a single mixture (Walsh, 2011). The Marshall stability is limited in order to ensure adequate deformation resistance whilst the water sensitivity is checked for surface course mixtures. The mixture is relatively dense which reduces the potential for foreign object damage, which is necessary for airfields with jet aircraft.

AC/MA can be used for all layers of the pavement with NMAS between 10 mm and 32 mm with the smallest sizes for regulating and the larger sizes for the lower layers. On airfields, the AC/MA is overlaid with a porous asphalt (PA) friction course in order to ensure adequate frictional properties in wet weather conditions. However, AC/MA is not always economic for small jobs when the material quantities do not warrant a dedicated mobile plant.

Generally, AC/MA has the following compositional aspects:

- Aggregate type – AC/MA requires no specific aggregate type other than high polished stone value for surface course.
- NMAS – AC/MA can be of any aggregate size.
- Aggregate grading – The grading for AC/MA is dense and tightly controlled.
- Binder content – The binder content of AC/MA is relatively low.
- Binder grade – The binder grade for AC/MA is moderate.
- Air voids content – AC/MA has a moderate air voids content.

The properties of AC/MA are typically:

- Resistance to deformation – Good.
- Resistance to cracking – Moderate.
- Permeability – Impermeable.
- Resistance to water damage – Moderate.
- Stiffness – Good.
- Resistance to fatigue – Moderate.
- Tensile strength – Good.
- Skid resistance – Poor, requiring grooving or friction course on airfields.
- Cohesion – Good.
- Tyre/road noise – Moderate.
- Spray generation – Poor.
- Resistance to chemical damage – Moderate.
- Adhesion – Moderate.
- Durability – Good.
- Cost – Relatively high.

4.2.3 Superpave Asphalt (AC/SP)

The Strategic Highway Research Program (SHRP) was set up in the USA to develop a method for designing asphalt mixtures in terms of performance properties rather than the more compositional design approaches of the Hveem and Marshall methods. The resulting method is called Superpave (Asphalt Institute, 2001; Cominsky et al., 1994), and the AC designed by the method can be termed Superpave asphalt (AC/SP^2). The properties of the mixture that are involved in the design method include tests for permanent deformation, fatigue cracking and low temperature cracking.

The tests involved depend on the three available levels (Level 1 by volumetrics, Level 2 for intermediate traffic levels and Level 3 for high traffic levels) to which the asphalt is developed.

The associated performance graded binder, also developed by SHRP, is based on providing resistance to rutting, fatigue cracking and low-temperature cracking at specific pavement temperatures. The tests used include dynamic shear rheometer, rotational viscometer, bending beam rheometer, direct tension tester, rolling thin film oven and pressure ageing vessel. The binder temperature ranges in the grading are based on the high and low temperatures at which a binder reaches critical values of distress-predicting properties.

AC/SP can be used almost universally, although it is generally used for more critical uses because of the additional aspects of the design and its associated cost. For the purposes of this book, the term 'Superpave asphalt' is used to imply an asphalt designed to the full (Level 3) method.

Generally, AC/SP has the following compositional aspects:

- Aggregate type – AC/SP requires no specific aggregate type other than high polished stone value for surface course.
- NMAS – AC/SP can be of any aggregate size.
- Aggregate grading – The grading for AC/SP is dense and tightly controlled.
- Binder content – The binder content of AC/SP is moderate.
- Binder grade – The binder grade for AC/SP is high grade of moderate viscosity and nearly always polymer modified.
- Air voids content – AC/SP has a moderate air voids content.

The properties of AC/SP are typically:

- Resistance to deformation – Good.
- Resistance to cracking – Moderate.
- Permeability – Impermeable.
- Resistance to water damage – Moderate.
- Stiffness – Good.
- Resistance to fatigue – Moderate.
- Tensile strength – Good.
- Skid resistance – Moderate.
- Cohesion – Good.
- Tyre/road noise – Moderate.
- Spray generation – Moderate.
- Resistance to chemical damage – Moderate.
- Adhesion – Moderate.
- Durability – Good.
- Cost – Relatively high.

4.2.4 Bitumen Macadam

4.2.4.1 General

The first macadam was developed in the UK by John Loudon McAdam around 1820 with regular sized stones that were relatively small for that time. The concept was enhanced by binding the aggregate particles with tar (hence the term 'tarmacadam' or 'tarmac' by which the general public in the UK call any asphalt). Subsequently, bitumen was used as the binder, initially as an alternative but subsequently as its replacement when tar was found to be carcinogenic (Walsh, 2011). With the production of a set of harmonised European standards, this category of mixtures is classified as a sub-category of AC but the name will be retained in this book for simplicity.

The gradings for macadams are generally taken from standard envelopes with PD 6691 (BSI, 2016) giving example specifications for the mixtures as widely used in the UK. Other European standardisation bodies may have example specifications in their guidance documents for similar mixtures. Macadams are generally not designed in terms of optimising the aggregate and/or binder although the mixtures can be checked for certain properties, in particular deformation resistance. Nevertheless, macadams still have an aggregate grading that is fairly closely defined. The binder contents of macadams tend to be relatively modest, generally around 4%. The low binder content helps to reduce the potential for deformation but does lead to reduced durability.

4.2.4.2 Dense Bitumen Macadam (AC/DBM)

AC/DBM are mixtures that have a Fuller index near 0.45 to 0.5 so as to be relatively dense and were traditionally made with 100/150 or 160/220 bitumen (with the latter being more commonly used in colder climates) but more recently with 40/60 bitumen, known as DBM50, in order to increase the strength of the mixture.

Generally, AC/DBM has the following compositional aspects:

- Aggregate type – AC/DBM requires no specific aggregate type other than high polished stone value for surface course.
- NMAS – AC/DBM can be of any aggregate size.
- Aggregate grading – The grading for AC/DBM is dense.
- Binder content – The binder content of AC/DBM is relatively low.
- Binder grade – The binder grade for AC/DBM is moderate.
- Air voids content – AC/DBM has a moderate air voids content.

The properties of AC/DBM are typically:

- Resistance to deformation – Moderate.
- Resistance to cracking – Moderate.
- Permeability – Impermeable.
- Resistance to water damage – Moderate.
- Stiffness – Good.
- Resistance to fatigue – Moderate.
- Tensile strength – Good.
- Skid resistance – Moderate.
- Cohesion – Moderate.
- Tyre/road noise – Moderate.
- Spray generation – Poor.
- Resistance to chemical damage – Moderate.
- Adhesion – Moderate.
- Durability – Moderate.
- Cost – Relatively low.

4.2.4.3 Heavy Duty Mixture (AC/HDM) and High Modulus Base (AC/HMB)

AC/HDM are mixtures that are similar to AC/DBM50 but with a higher filler content in order to further increase the strength of the mixture. AC/HMB are mixtures that are similar to AC/HDM but with 30/45 bitumen in order to increase the strength of the mixture even further.

Generally, AC/HDM and AC/HMB have the following compositional aspects:

- Aggregate type – AC/HDM and AC/HMB require no specific aggregate type other than high strength.
- NMAS – AC/HDM and AC/HMB can be of any aggregate size but generally of the medium and larger sizes because the mixtures are only used in the lower layers.
- Aggregate grading – The gradings for AC/HDM and AC/HMB are dense.
- Binder content – The binder contents of AC/HDM and AC/HMB are moderate.
- Binder grade – The binder grades for AC/HDM and AC/HMB are moderate.
- Air voids content – AC/HDM and AC/HMB have moderate air voids contents.

The properties of AC/HDM and AC/HMB are typically:

- Resistance to deformation – Good.
- Resistance to cracking – Moderate.
- Permeability – Impermeable.

- Resistance to water damage – Moderate.
- Stiffness – Good.
- Resistance to fatigue – Moderate.
- Tensile strength – Good.
- Skid resistance – Not relevant.
- Cohesion – Moderate.
- Tyre/road noise – Not relevant.
- Spray generation – not relevant.
- Resistance to chemical damage – Moderate.
- Adhesion – Moderate.
- Durability – Moderate.
- Cost – Moderate.

4.2.4.4 More Open Bitumen Macadams (AC/OBM)

There are other macadam mixtures that are more open than DBM (AC/OBM[3]), the use of which is generally restricted to the surface course on local roads. The more open structure is achieved by less fines, which also allows a reduced binder content and, hence, reduced cost with the reduced surface area. The more open mixtures given in PD 6691 (British Standard Institution, 2016) are AC 20 open bin, AC 14 open surf, AC 10 open surf, AC 14 close surf, AC 10 close surf, AC 6 dense surf, AC 6 med surf and AC 4 fine surf. The classifications 'close', 'med' (medium) and 'open' indicate increasing degrees of openness, with the 'open' mixtures becoming almost dense PA (Section 4.6). The fine graded mixture, AC 4 fine surf, has an NMAS of 4 mm and is generally used for regulating purposes.

AC/OBM has been used widely for all layers on local roads because of their relatively low cost resulting from the low binder content, the most expensive component in asphalt. In addition, the inherent strength of the mixtures, particularly AC/DBM50, AC/HDM and AC/HMB, also makes them a good material for the structural layers for all pavements. However, the low binder content also limits the durability of the material. It has not been used regularly on heavily trafficked roads in the surface course, for which the main causes for deterioration are the environment and the direct action of the traffic. The deterioration tends to be localised fretting and general deterioration that gets worse gradually with time, thus giving warning of the end of its serviceability and allowing time to plan its replacement.

Generally, AC/OBM has the following compositional aspects:

- Aggregate type – AC/OBM requires no specific aggregate type.
- NMAS – AC/OBM can be of any aggregate size.
- Aggregate grading – The grading for AC/OBM is open.

- Binder content – The binder content of AC/OBM is low.
- Binder grade – The binder grade for AC/OBM is moderate.
- Air voids content – AC/OBM has a relatively high air voids content.

The properties of AC/OBM are typically:

- Resistance to deformation – Moderate.
- Resistance to cracking – Moderate.
- Permeability – Relatively permeable.
- Resistance to water damage – Relatively poor.
- Stiffness – Moderate.
- Resistance to fatigue – Relatively poor.
- Tensile strength – Moderate.
- Skid resistance – Moderate.
- Cohesion – Moderate.
- Tyre/road noise – Moderate.
- Spray generation – Relatively good.
- Resistance to chemical damage – Relatively poor.
- Adhesion – Moderate.
- Durability – Relatively poor.
- Cost – Low.

4.2.5 Enrobés À Module Élevé (AC/EME)

Enrobé à module élevé (AC/EME) (Figure 4.4) is an AC where the binder content has been increased to enhance the durability whilst the stiffness of the bitumen used has been increased in order to counter the reduction in strength and deformation resistance that result from an increased binder content. The mixture was developed in France with two categories, EME1 and EME2, of which EME2 has the higher binder content. A modified version of EME1 was introduced into the UK as high modulus base mixture with 15 pen bitumen. However, HBM15 proved not to be durable and, therefore, was withdrawn. Subsequently, EME2 was introduced into the UK but without significant changes from the original design (Sanders and Nunn, 2005).

PD 6691 (British Standard Institution, 2016) includes an example specification for AC/EME2 for use in the UK and other European standardisation bodies may have similar example specifications in their guidance documents. AC/EME needs to be designed carefully against several properties because of the high performance expected of the mixture.

AC/EME2 is used for base and binder course layers; the binder is too stiff to be used in the surface course. As a result of this stiff binder, the mixture is also very stiff and, therefore, can allow a reduced pavement thickness for the same service life, or an extended service life for the same thickness. However, because of its stiffness, a strong substrate foundation

Figure 4.4 AC/EME2 after light site trafficking.

is required, which limits the sites where it can be used. The NMASs for AC/EME2 in the UK are 10 mm, 14 mm and 20 mm.

Generally, AC/EME has the following compositional aspects:

- Aggregate type – AC/EME requires no specific aggregate type other than high strength.
- NMAS – AC/EME can be of any aggregate size but generally of the medium to larger sizes because the mixture is only used in the lower layers.
- Aggregate grading – The grading for AC/EME is dense.
- Binder content – The binder content of AC/EME is moderate.
- Binder grade – The binder grade for AC/EME is very stiff.
- Air voids content – AC/EME has a relatively low air voids content.

The properties of AC/EME are typically:

- Resistance to deformation – Good.
- Resistance to cracking – Moderate.
- Permeability – Impermeable.

- Resistance to water damage – Good.
- Stiffness – Very good.
- Resistance to fatigue – Moderate.
- Tensile strength – Good.
- Skid resistance – Not relevant.
- Cohesion – Moderate.
- Tyre/road noise – Not relevant.
- Spray generation – Not relevant.
- Resistance to chemical damage – Moderate.
- Adhesion – Very poor.
- Durability – Good.
- Cost – Relatively high.

4.2.6 Béton Bitumineux Pour Chaussées Aéronautiques (AC/BBA)

Béton bitumineux pour chaussées aéronautiques (AC/BBA) is a form of AC that was developed in France for airfield pavements, where it has been used for many years. More recently, the material has been used in the UK with an example specification being included in PD 6691 (British Standard Institution, 2016); other European standardisation bodies may have similar example specifications in their guidance documents.

There are four categories of BBA material with NMAS of either 10 mm or 14 mm with a close (C) or gap/discontinuous (D) grading for both sizes. For type C mixtures, the aggregate grading curve is smooth and continuous while, for type D mixtures, the discontinuity in the grading is formed by omitting either the 4/6 mm or the 2/6 mm aggregate fraction. The binder is generally unmodified bitumen (70/100, 50/70, 35/50, 20/30 or 10/20 pen) although polymer-modified bitumen (PMB) is used for specific requirements.

BBA can be used for binder and surface courses in new construction and overlay. The main features of BBA are high durability, good resistance to rutting and good resistance to punching and cracking. These properties make BBA well suited for pavement runways heavily loaded with channellised heavy traffic as well as slow lanes, ramps and bus lanes of highways and industrial platforms.

Generally, BBA has the following compositional aspects:

- Aggregate type – BBA requires no specific aggregate type other than high polished stone value for surface course and high strength.
- NMAS – BBA can be of any aggregate size.
- Aggregate grading – The grading for BBA is dense.
- Binder content – The binder content of BBA is moderate.
- Binder grade – The binder grade for BBA is moderate.
- Air voids content – BBA has a relatively low air voids content.

The properties of BBA are typically:

- Resistance to deformation – Good.
- Resistance to cracking – Moderate.
- Permeability – Impermeable.
- Resistance to water damage – Moderate.
- Stiffness – Good.
- Resistance to fatigue – Moderate.
- Tensile strength – Good.
- Skid resistance – Moderate.
- Cohesion – Good.
- Tyre/road noise – Moderate.
- Spray generation – Moderate.
- Resistance to chemical damage – Moderate.
- Adhesion – Moderate.
- Durability – Relatively good.
- Cost – Relatively high.

4.3 Stone Mastic Asphalt (SMA)

SMA is an asphalt mixture that is standardised in EN 13108-5 (CEN, 2016b) that is based on aggregate interlock to maximise its deformation resistance and with the remaining voids filled with mortar (Figure 4.5). It was developed in Germany as a composite between AC (Section 4.2) and mastic asphalt (MA) (Section 4.7) in order to overcome abrasion by studded tyres, subsequently solved by banning those tyres. The mortar requires a higher binder content than the aggregate skeleton can take without draining, and fibres are included to increase the available surface area and, hence, avoid binder drainage. PMB can be used instead or as well for this purpose, although PMB can also be used to enhance other properties.

The air voids content of SMA should be relatively low, although it was increased to provide higher texture depth when introduced into the UK as proprietary TSCS. However, the higher air voids content does come at a cost in terms of durability (Nicholls et al., 2010).

SMA can be used for all pavement layers, although it is generally used for the surface and binder courses. It is also used for most types of paved areas in new construction and maintenance.

Generally, SMA has the following compositional aspects:

- Aggregate type – SMA requires no specific aggregate type other than high polished stone value for surface course.
- NMAS – SMA can be of any aggregate size.

Figure 4.5 Freshly laid SMA.

- Aggregate grading – The grading for SMA is gap graded.
- Binder content – The binder content of SMA is relatively high.
- Binder grade – The binder grade for SMA is moderate.
- Air voids content – SMA has a moderate air voids content.

The properties of SMA are typically:

- Resistance to deformation – Good.
- Resistance to cracking – Moderate.
- Permeability – Relatively impermeable.
- Resistance to water damage – Moderate.
- Stiffness – Good.
- Resistance to fatigue – Moderate.
- Tensile strength – Moderate.
- Skid resistance – Relatively good.
- Cohesion – Good.
- Tyre/road noise – Good.
- Spray generation – Relatively good.
- Resistance to chemical damage – Moderate.
- Adhesion – Moderate.
- Durability – Moderate.
- Cost – Moderate.

4.4 Asphalt Concrete for Very Thin Layers (BBTM)

BBTM[4] as shown in Figure 4.6, is an asphalt mixture that is standardised in EN 13108-2 (CEN, 2016c). BBTM was developed in France and subsequently introduced into other European states, first coming to the UK in 1992 as proprietary TSCS.

The concept for BBTM is based on a modification to the AC aggregate distribution that makes it gap graded such that it can be laid at thicknesses that are relatively thin for the maximum aggregate size. However, the mixtures require a PMB in order to provide a thick binder film on the aggregates without binder drainage that will result in the required strength and durability expected of the material.

BBTM is used as a surface course material on all categories of roads and usually deteriorates by fretting and/or surface cracking. However, the rate of loss of aggregate particles often accelerates after the deterioration has started if the binder has aged to become brittle from the lack of support from previously adjacent particles being missing. The use of PMB can help to extend the time before the binder becomes too brittle.

Some BBTM mixtures are not easy to hand-lay and should, therefore, not be used on more intricate sites, although there are no problems when BBTM is machine-laid. In the UK, BBTM can be relatively permeable because the mixtures have been opened up from the original

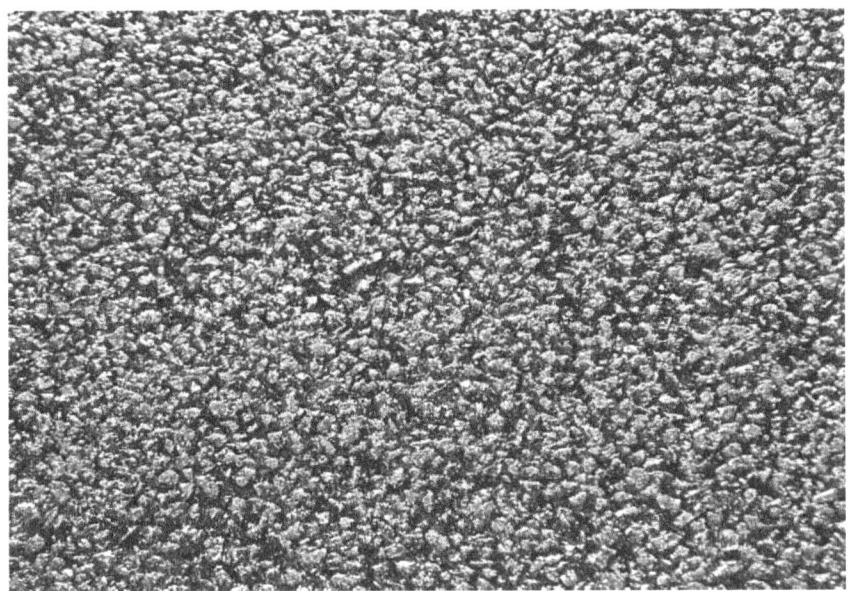

Figure 4.6 Freshly laid BBTM.

designs in order to achieve the UK texture requirements. These mixtures should not be used in areas where water could be a problem and such mixtures can be avoided by setting an upper limit on the air voids content, although such a limit will inhibit high texture depths without using a high NMAS.

Generally, BBTM has the following compositional aspects:

- Aggregate type – BBTM requires no specific aggregate type other than high polished stone value.
- NMAS – BBTM can be of any aggregate size.
- Aggregate grading – The grading for BBTM is gap graded.
- Binder content – The binder content of BBTM is relatively high.
- Binder grade – The binder grade for BBTM is polymer modified.
- Air voids content – BBTM has a moderate air voids content.

The properties of BBTM are typically:

- Resistance to deformation – Good.
- Resistance to cracking – Moderate.
- Permeability – Relatively impermeable.
- Resistance to water damage – Moderate.
- Stiffness – Good.
- Resistance to fatigue – Moderate.
- Tensile strength – Moderate.
- Skid resistance – Relatively good.
- Cohesion – Good.
- Tyre/road noise – Good.
- Spray generation – Relatively good.
- Resistance to chemical damage – Moderate.
- Adhesion – Moderate.
- Durability – Moderate.
- Cost – Moderate.

4.5 Asphalt for Ultra-Thin Layers (AUTL)

AUTL is an asphalt mixture that is standardised in EN 13108-9 (CEN, 2016d). AUTL (Figure 4.7) is spread using a purpose-built machine that incorporates a binder sprayer to spray a thick bond coat immediately before the asphalt is laid. The long length of the machine enables an excellent longitudinal profile to be achieved. The sprayed binder is a modified emulsion, containing approximately 70% solids, and is sprayed at a rate of about 1 l/m^2. The binder in the actual mixture is usually paving grade whilst the aggregate, which is close to being single sized, if not to the same

Figure 4.7 Freshly laid AUTL.

extent as PA (Section 4.6), has an NMAS of 14 mm, 10 mm or 6 mm. Because of the single size, it can be laid at little over the aggregate size and has also been called paver-laid surface dressing.

AUTL is generally only used for surface courses of roads.

Generally, AUTL has the following compositional aspects:

- Aggregate type – AUTL requires no specific aggregate type other than high polished stone value.
- NMAS – AUTL can be of any aggregate size.
- Aggregate grading – The grading for AUTL is gap graded.
- Binder content – The binder content of AUTL is relatively high.
- Binder grade – The binder grade for AUTL is moderate.
- Air voids content – AUTL has a moderate air voids content.

The properties of AUTL are typically:

- Resistance to deformation – Good.
- Resistance to cracking – Moderate.
- Permeability – Relatively impermeable.
- Resistance to water damage – Moderate.
- Stiffness – Good.

- Resistance to fatigue – Moderate.
- Tensile strength – Moderate.
- Skid resistance – Relatively good.
- Cohesion – Good.
- Tyre/road noise – Good.
- Spray generation – Relatively good.
- Resistance to chemical damage – Moderate.
- Adhesion – Moderate.
- Durability – Moderate.
- Cost – Moderate.

4.6 Porous Asphalt (PA)

4.6.1 Conventional Porous Asphalt (PA)

PA is an asphalt mixture that is standardised in EN 13108-7 (CEN, 2016e). PA (Figure 4.8) has an almost single size aggregate in order to maximise the air voids content, which is generally around 20%. The binder film needs to be relatively thick in order to limit the binder oxidation in such an open structure. Generally, the amount of binder required is greater than will remain on the aggregate particles and either fibres or

Figure 4.8 Trafficked PA.

PMB are needed to avoid binder drainage. The NMAS that was found to be most efficient for roads in the UK was 20 mm (Nicholls, 1997) whilst the size used on airfields tends to be 10 mm, which is more similar to the size used on roads in other European countries.

The usual failure mechanism is fretting. Whilst in early life the binder will hold the remaining particles in place, the fretting becomes catastrophic once the binder becomes brittle. The point at which the binder becomes brittle is about 15 pen for paving grade binders (Nicholls, 1997), which allows the date to be estimated if the penetration of recovered binder is routinely monitored.

Another failure mechanism (in terms of providing the desired properties of PA rather than its structural integrity) is the clogging up of the pores with detritus. The clogging up can be reduced by cleaning using suction, although it is only a partial remedy.

PA was originally developed to minimise aquaplaning on runways, for which it is still used, when it is called porous friction course. The mixture was then trialled for its spray-reducing properties on roads but was taken up for its noise-reduction properties, particularly in Belgium and the Netherlands. When PA is used as the surface course of a pavement, there needs to be drainage from the top of the highest non-porous layer for when the voids are open and from the surface for when the voids are clogged. The highest non-porous layer also needs to be impermeable because water can be held unseen at the bottom of the PA.

As well as use for road and airfield surface courses, PA is also used in sustainable drainage systems (SuDS). SuDS aim to alleviate the problems of excess water in urban areas with extensive areas of impermeable surfacing by storing and/or slowing down the flow of surface water. For SuDS, which are generally used for car parks and other extended lightly trafficked areas, the full depth of the pavement needs to be in PA and drainage needs to be arranged to take the water away from the bottom of the pavement. Because the water will pass through the pavement, the unbound layers have to be designed to take water passing through them. The stiffness of PA is generally lower than that of other asphalt types because of its open structure and, as such, is not generally used in other structural layers, particularly in heavily trafficked pavements.

Generally, PA has the following compositional aspects:

- Aggregate type – PA requires high polished stone value for surface course and high strength.
- NMAS – PA cannot be of a very small aggregate size.
- Aggregate grading – The grading for PA is practically single sized and very open.
- Binder content – The binder content of PA is limited by the need to avoid binder drainage and avoid filling the voids.

- Binder grade – The binder grade for PA is moderate with polymer modification to reduce binder drainage.
- Air voids content – PA has a very high air voids content.

The properties of PA are typically:

- Resistance to deformation – Good.
- Resistance to cracking – Moderate.
- Permeability – Permeable with large interconnected voids.
- Resistance to water damage – Moderate.
- Stiffness – Low.
- Resistance to fatigue – Moderate.
- Tensile strength – Moderate.
- Skid resistance – Good.
- Cohesion – Good.
- Tyre/road noise – Very good.
- Spray generation – Good.
- Resistance to chemical damage – Poor.
- Adhesion – Moderate.
- Durability – Relatively poor.
- Cost – Relatively high.

4.6.2 Twin-Layer Porous Asphalt (PA/TL)

PA/TL[5] is a composite mixture for surface courses with a smaller NMAS PA above a larger NMAS PA so that the detritus that is so detrimental to the spray- and noise-reducing properties is collected in the top layer. Cleaning by suction of the top layer can then be undertaken more easily because of the open layer underneath the clogged upper layer. PA/TL is treated as a combined surface layer because it will still need an appropriate binder course underneath it to control the water.

Generally, twin-layer porous asphalt has the following compositional aspects:

- Aggregate type – PA/TL requires no specific aggregate type other than high polished stone value for the top layer and high strength.
- NMAS – The bottom layer is made with a larger sized aggregate and the top layer with smaller size.
- Aggregate grading – The gradings of both layers are practically single sized and very open.
- Binder content – The binder contents of both layers are limited by the need to avoid binder drainage and fill the voids.
- Binder grade – The binder grades for both layers are moderate with polymer modification to reduce binder drainage.

- Air voids content – Both layers have very high air voids contents.

The properties of PA/TL are typically:

- Resistance to deformation – Good.
- Resistance to cracking – Moderate.
- Permeability – Permeable.
- Resistance to water damage – Moderate.
- Stiffness – Low.
- Resistance to fatigue – Moderate.
- Tensile strength – Moderate.
- Skid resistance – Good.
- Cohesion – Good.
- Tyre/road noise – Very good.
- Spray generation – Good.
- Resistance to chemical damage – Poor.
- Adhesion – Moderate.
- Durability – Relatively poor.
- Cost – Relatively high.

4.6.3 Grouted Macadam (PA/GM)

PA/GM[6] is a composite material in which the voids in an open porous asphalt are filled with a cementitious grout. The highly permeable material is, therefore, converted into a highly impermeable material. It is used for heavily loaded areas and locally on PA surface courses in areas of high stress such as roundabouts.

Generally, PA/GM has the following compositional aspects:

- Aggregate type – Asphalt requires no specific aggregate type other than high polished stone value for surface course.
- NMAS – PA/GMs are generally made with larger aggregate sizes.
- Aggregate grading – The grading of PA/GM is practically single sized and open prior to grouting.
- Binder content – The binder content of PA/GM is moderate.
- Binder grade – The binder grade for PA/GM is moderate.
- Air voids content – PA/GM has a high air voids content prior to grouting and a low air voids content afterwards.

The properties of PA/GM are typically:

- Resistance to deformation – Good.
- Resistance to cracking – Good.
- Permeability – Impermeable once grouted.

- Resistance to water damage – Good.
- Stiffness – Moderate.
- Resistance to fatigue – Relatively good.
- Tensile strength – Moderate.
- Skid resistance – Moderate.
- Cohesion – Good.
- Tyre/road noise – Moderate.
- Spray generation – Moderate.
- Resistance to chemical damage – Good.
- Adhesion – Moderate.
- Durability – Good.
- Cost – Relatively high.

4.7 Mastic Asphalt (MA)

4.7.1 General

MA is an asphalt mixture that is standardised in EN 13,108-6 (Comité Européen de Normalisation, 2016e). MA is a mortar-based mixture with a maximum fine aggregate size of 2 mm and 10/14 mm coarse aggregate particles embedded as 'plums' and 8/14 mm or 14/20 mm pre-coated chippings when used as the surface course. The bitumen, which is added at between 30% and 50% of the mixture, has to be very stiff at between 13 pen and 25 pen. MA is fluid enough that it has to be hand laid and screeded off rather than compacted (Walsh, 2011).

With the very high binder content, MA is the most impermeable of the asphalt types and, with the stiff binder, reasonably deformation-resistant, although using an indentation test rather than wheel-tracking, partly to demonstrate resistance to high heels for when the material is used on footways.

The primary uses of MA are in tunnels, concrete bridge decks, steel bridge decks, footpaths and flat roofs, where there is a need for a water-resistant material that can be laid thinly (because of the small aggregate size) and with a smooth finish. However, the high binder content makes the material relatively expensive and not widely used outside these specialist areas.

Generally, MA has the following compositional aspects:

- Aggregate type – MA requires no specific aggregate type other than high polished stone value for the pre-coated chippings.
- NMAS – MA is made with smaller aggregate sizes.
- Aggregate grading – The grading of MA is dense.
- Binder content – The binder content of MA is high.
- Binder grade – The binder grade for MA is stiff.
- Air voids content – MA has a very low air voids content.

The properties of MA are typically:

- Resistance to deformation – Moderate.
- Resistance to cracking – Good.
- Permeability – Impermeable.
- Resistance to water damage – Good.
- Stiffness – Relatively low.
- Resistance to fatigue – Good.
- Tensile strength – Moderate.
- Skid resistance – Moderate.
- Cohesion – Good.
- Tyre/road noise – Moderate.
- Spray generation – Relatively poor.
- Resistance to chemical damage – Good.
- Adhesion – Good.
- Durability – Good.
- Cost – Relatively high.

4.7.2 Gussaphalt (MA/Guss)

There is an MA variant called gussasphalt (MA/Guss[7]) (which can be translated as 'poured asphalt') that can be machine laid. MA/Gus uses a coarser fine aggregate and more binder (up to 75%) that is less stiff (20 pen to 30 pen). It is used in similar situations as the hand-laid MA.

The compositional aspects and properties of MA/Gus are the same as for MA.

4.8 Hot Rolled Asphalt (HRA)

4.8.1 General

HRA is an asphalt mixture that is standardised in EN 13108-4 (CEN, 2016f). HRA is a mortar-based mixture with coarse aggregate particles inserted, originally to reduce cost (given aggregate is less expensive than binder by volume) but subsequently to improve the stiffness and deformation resistance. With the gap-grading needed in the particle distribution, the aggregate envelopes of different HRA mixtures are separate and there is no overall envelope in EN 13108-4.

HRA has limited aggregate interlock, as shown in Figure 4.9, when compared to SMA (Section 4.3), and the strength of the mixtures is primarily derived from the strength of the mortar. Therefore, the mechanical properties are much more temperature susceptible than those of mixture based on aggregate interlock, which could be a worsening problem with

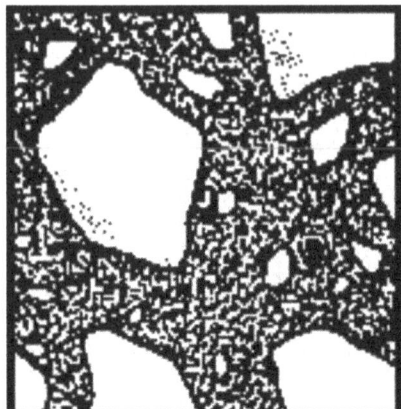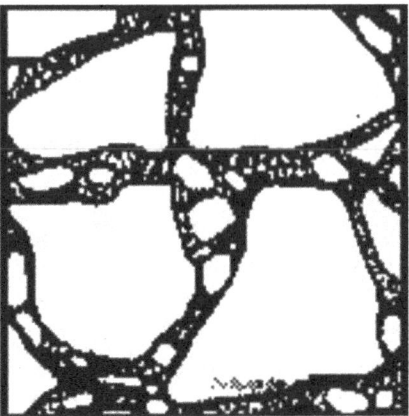

Figure 4.9 Coarse aggregate in HRA (left) and SMA (right).

global warming. However, the binder stiffness will increase with ageing, reducing the potential to deform.

Because they are mortar-based, HRA mixtures are very impermeable and good when there is a need to protect lower pavement layers. The binder used for HRA is relatively stiff, with 40/60 pen binder typically being used in the UK, which was the principal country in Europe in terms of the use of HRA. That particular grade was only included in the CEN slate of grades at the insistence of the UK. HRA can be designed using the Marshall method, optimising the binder content for stability, flow and certain other properties. A check on the deformation resistance using wheel-tracking has been added, often requiring the use of a PMB. However, many HRA mixtures are recipe-based rather than designed (Nicholls, 1998; Walsh, 2011).

4.8.2 Hot Rolled Asphalt with Pre-Coated Chippings (HRA+PCC)

HRA with pre-coated chippings (PCC) are surface course mixtures. The proportion of coarse aggregate is relatively low, generally at either 30% or 35%, with an NMAS of 10 m or 14 mm. The PCC is generally the fraction above the NMAS of the HRA, that is, 8/14 mm or 14/20 mm. The PCC are sprinkled on the HRA after laying but before compaction to be rolled in to provide skid-resistance. The protruding extremities of the PCC provide the texture depth while the aggregate source can be selected to provide the micro-texture required for skid resistance without all the surface course having to have the same properties.

HRA with PCC (Figure 4.10) can fail prematurely when the chippings are not adequately embedded and quickly become loose. However, deterioration is usually gradual with plenty of warning. Two warning signs are the loss of chippings and the emergence of cracks.

Generally, HRA with PCC has the following compositional aspects:

- Aggregate type – HRA with PCC requires no specific aggregate type other than high polished stone value for the pre-coated chippings.
- NMAS – NMAS of HRA is generally smaller than aggregate size of the PCC.
- Aggregate grading – The grading of HRA is gap grade with continuous smaller sizes with larger sizes inserted as 'plumbs'.
- Binder content – The binder content of HRA is relatively high.
- Binder grade – The binder grade for HRA is moderate with polymer-modification used to improve deformation and chipping retention.
- Air voids content – HRA has a relatively low air voids content.

Figure 4.10 HRA with pre-coated chippings after 16 years in service.

The properties of HRA + PCC are typically:

- Resistance to deformation – Relatively good with polymer modification.
- Resistance to cracking – Good.
- Permeability – Impermeable.
- Resistance to water damage – Good.
- Stiffness – Moderate.
- Resistance to fatigue – Moderate.
- Tensile strength – Moderate.
- Skid resistance – Good.
- Cohesion – Good.
- Tyre/road noise – Moderate.
- Spray generation – Moderate.
- Resistance to chemical damage – Relatively good.
- Adhesion – Good.
- Durability – Relatively good.
- Cost – Relatively high.

4.8.3 High Stone-Content Hot Rolled Asphalt (HRA/HSC)

HRA with a higher proportion of coarse aggregate (HRA/HSC[8]), generally 50% or 55%, without any PCC can be used for any of the pavement layers. HRA/HSC generally has an NMAS of 10 mm, 14 mm, 20 mm or 32 mm.

HRA/HSC used for the surface course takes time to develop its skid-resistant properties which occurs from fretting of the mortar from between the coarse particles on the surface. As such, it is not suitable for heavily trafficked sites or other situations requiring good skid resistance. On lesser trafficked situations, the time taken to achieve skid resistance will be greater because of the limited traffic to scuff away the mortar.

HRA/HSC has been widely used as the structural layers of the pavement in the UK. The loss of strength with elevated temperatures will be less important for these layers than the surface course with reduced temperature variability at depth.

Generally, HRA/HSC has the following compositional aspects:

- Aggregate type – HRA/HSC requires no specific aggregate type other than high polished stone value for surface course.
- NMAS – HRA/HSC can be of any aggregate size.
- Aggregate grading – The grading of HRA/HSC is gap grade with continuous smaller sizes with larger sizes inserted as 'plumbs'.
- Binder content – The binder content of HRA/HSC is relatively high.
- Binder grade – The binder grade for HRA is moderate.
- Air voids content – HRA has a relatively low air voids content.

The properties of HRA/HSC are typically:

- Resistance to deformation – Relatively good with polymer modification.
- Resistance to cracking – Good.
- Permeability – Impermeable.
- Resistance to water damage – Good.
- Stiffness – Moderate.
- Resistance to fatigue – Moderate.
- Tensile strength – Moderate.
- Skid resistance – Good once coarse aggregate exposed.
- Cohesion – Good.
- Tyre/road noise – Moderate.
- Spray generation – Moderate.
- Resistance to chemical damage – Relatively good.
- Adhesion – Good.
- Durability – Relatively good.
- Cost – Moderate.

4.8.4 Sand Carpet (HRA/SC)

A further type of HRA without any coarse aggregate (sand carpet, HRA/SC[9]) has been used as a protective layer above bridge-deck waterproofing, often with a red pigment to act as a warning during remedial works. However, the lack of coarse aggregate means that HRA/SC is not very resistant to deformation and the red pigment fades with time. As such, the use of HRA/SC has been widely replaced by other mixtures with low NMAS to avoid puncturing the waterproofing sheet.

Generally, HRA/SC has the following compositional aspects:

- Aggregate type – HRA/SC requires no specific aggregate type.
- NMAS – HRA/SC uses small aggregate sizes.
- Aggregate grading – The grading of HRA/SC is gap graded.
- Binder content – The binder content of HRA/SC is relatively high.
- Binder grade – The binder grade for HRA/SC is moderate.
- Air voids content – HRA/SC has a relatively low air voids content.

The properties of HRA/SC are typically:

- Resistance to deformation – Poor.
- Resistance to cracking – Moderate.
- Permeability – Impermeable.
- Resistance to water damage – Moderate.
- Stiffness – Poor.
- Resistance to fatigue – Moderate.

- Tensile strength – Moderate.
- Skid resistance – Not relevant.
- Cohesion – Moderate.
- Tyre/road noise – Not relevant.
- Spray generation – Not relevant.
- Resistance to chemical damage – Moderate.
- Adhesion – Moderate.
- Durability – Relatively poor.
- Cost – Moderate.

4.9 Soft Asphalt (SA)

SA is an asphalt mixture that is commonly used as a paving material in Scandinavia and is standardised in EN 13108-3 (Comité Européen de Normalisation, 2016h). As the name suggests, the mixtures use relative soft binders of pen grades 250/330 up to 650/900 and soft bitumen grade from V1500 up to V12000. The soft binders not only make the mixture easy to lay in the cold conditions that occur in Scandinavia but also provide flexibility to withstand deformations and temperature fluctuations in the area with a balance between the flexibility and the stiffness. SA differs from other asphalt mixtures with low temperature mixing, greater elasticity, greater durability, less sensitivity to frost heave and to fatigue and is capable of self-healing (Vaitkus et al., 2016). The self-healing results from the bitumen becoming significantly softer during warm weather. However, SA is not particularly stable and has limited resistance to abrasion, which is particularly relevant in Scandinavia with studded tyres.

There are four types of SA (Type A, Type B, Type C and Type S) with the type being chosen according to desirable mechanical properties under climatic conditions. Each type contains both open and dense surface course options with NMAS between 10 mm and 14 mm while Type S also includes dense base options with 14 mm and 20 mm NMAS.

SA roads constitute a relatively large and important part of the lightly trafficked road network in Scandinavia with more than 10,000 km in Sweden. A single surface course layer of SA can be used to overlay gravel roads with traffic of up to 500 vehicles per day. The deterioration of such roads is generally from climatic factors such as low-temperature cracking and uneven deformation due to frost heave and thawing.

Generally, SA has the following compositional aspects:

- Aggregate type – SA requires no specific aggregate type.
- NMAS – SA can be of any aggregate size.
- Aggregate grading – The grading of SA is continuous.
- Binder content – The binder content of SA is moderate.

- Binder grade – The binder grade for SA is soft.
- Air voids content – SA has a relatively low air voids content.

The properties of SA are typically:

- Resistance to deformation – Moderate.
- Resistance to cracking – Good.
- Permeability – Impermeable.
- Resistance to water damage – Moderate.
- Stiffness – Moderate.
- Resistance to fatigue – Good.
- Tensile strength – Moderate.
- Skid resistance – Moderate.
- Cohesion – Good.
- Tyre/road noise – Moderate.
- Spray generation – Moderate.
- Resistance to chemical damage – Moderate.
- Adhesion – Good.
- Durability – Relatively poor.
- Cost – Relatively low.

Notes

1 Not a widely used acronym but one that is used for convenience in this book.
2 Ibid.
3 Ibid.
4 The acronym is based on the French, bétons bitumineux très minces.
5 Not a widely used acronym but one that is used for convenience in this book.
6 Ibid.
7 Ibid.
8 Ibid.
9 Ibid.

References

Asphalt Institute (2001). Superpave mix design. *Superpave Series No. 2 (SP02).* Lexington, KY: Asphalt Institute.

British Standard Institution (2016). Guidance on the use of BS EN 13108 "Bituminous mixtures – Material Specifications". *PD 6691:2015+A1:2016.* London: British Standard Institution.

Cominsky, R J, G A Huber, T W Kennedy, and M Anderson (1994). The Superpave mix design manual for new construction and overlays. SHRP-A-407. Washington, DC: Strategic Highway Research Program, National Research Council.

Comité Européen de Normalisation (2016a). Bituminous mixtures – Material specifications – Part 1, Asphalt concrete. *EN 13108-1:2016.* Most European standardisation institutions.

Comité Européen de Normalisation (2016b). Bituminous mixtures – Material specifications – Part 5, Stone mastic asphalt. *EN 13108-5:2016*. Most European standardisation institutions.

Comité Européen de Normalisation (2016c). Bituminous mixtures – Material specifications – Part 2, Asphalt concrete for very thin layers (BBTM). *EN 13108-2:2016*. Most European standardisation institutions.

Comité Européen de Normalisation (2016c). Bituminous mixtures – Material specifications – Part 9, Asphalt for ultra-thin layer (AUTL). *EN 13108-9:2016*. Most European standardisation institutions.

Comité Européen de Normalisation (2016d). Bituminous mixtures – Material specifications – Part 7, Porous asphalt. *EN 13108-7:2016*. Most European standardisation institutions.

Comité Européen de Normalisation (2016e). Bituminous mixtures – Material specifications – Part 6, Mastic asphalt. *EN 13108-6:2016*. Most European standardisation institutions.

Comité Européen de Normalisation (2016f). Bituminous mixtures – Material specifications – Part 4, Hot rolled asphalt. *EN 13108-4:2016*. Most European standardisation institutions.

Comité Européen de Normalisation (2016h). Bituminous mixtures – Material specifications – Part 3, Soft asphalt. *EN 13108-3:2016*. Most European standardisation institutions.

Defence Estates, Ministry of Defence (2009). Marshal asphalt for airfields. *SPEC 13* www.gov.uk/government/publications/specifications.

Fuller, W B, and S E Thompson (1907). The laws of proportioning concrete. In *Transactions of the American Society of Civil Engineers 59*, pp. 67–172. Reston, VA: American Society of Civil Engineers.

Nicholls, J C (1997). Review of UK porous asphalt trials. *TRL Report TRL264*. Wokingham: TRL Limited.

Nicholls, J C (editor) (1998). *Asphalt Surfacings*. London: CRC Press.

Nicholls, J C, I Carswell, C Thomas, and B Sexton (2010). Durability of thin asphalt surfacing systems: Part 4, final report after nine years' monitoring. *TRL Report TRL674*. Wokingham: TRL Limited.

Sanders, P J, and M E Nunn (2005). The application of enrobe à module élevé in flexible pavements. *TRL Report TRL636*. Wokingham: TRL Limited.

Vaitkus, A, V Vorobjovas, F Tuminiené, and J Gražulyté (2016). Experience in rehabilitation of low-volume roads using soft asphalt and Otta seal technologies. *6th Transport Research Arena, April 18-21, 2016*. https://ac.els-cdn.com/S2352146516302988/1-s2.0-S2352146516302988-main.pdf?_tid=cb81c6e1-ab9c-4522-9969-76a1eb42abae&acdnat=1529763498_88f0a10a7babc6a248aa44fbf9df72ce

Walsh, I D (editor) (2011). *ICE Manual of Highway Design and Management*. London: Thomas Telford.

Chapter 5

Paved Areas

5.1 General

This chapter reviews the principal areas that could require the use of asphalt. It will discuss the considerations that will need to be taken into account when selecting the materials for the relevant layers. It will be seen that in many of the applications there are conflicting and varying requirements to be considered as in these applications the asphalt may need to cope with varying conditions in terms of traffic loading, type and speed.

5.2 Roads

5.2.1 Motorways and Strategic Dual Carriageways

These roads (Figure 5.1) are the most heavily trafficked roads on the network both in terms of volume of traffic as well as the traffic loading applied and, while they only represent about 15% of the overall length, they take over 85% of the traffic. They are generally high-speed roads although many of the dual carriageways pass through urban areas where they will be subject to lower speed limits potentially down to 30 mph. Increasingly, the speed limits are being reduced on the motorways in the UK with the introduction of Smart Motorways where the speed limit is adjusted to take account of the traffic flows and can regularly be reduced to 40 or 50 mph or even less. A further consideration on motorways with at least three lanes is that the outside lane is usually restricted to cars only so does not have the traffic loading of the other lanes. This difference makes the selection of the layer thickness together with the choice of material a complicated matter.

The properties generally required of motorways and strategic dual carriageways are given in Table 5.1.

Figure 5.1 Typical motorway.

5.2.2 Urban Roads

These roads (Figure 5.2) are generally not the heaviest trafficked roads, but they do carry a varied mix of vehicular traffic. This variation can be roads which predominantly take all cars to roads which are used by a large amount of heavy goods vehicles. The speed limit on these roads is also variable and can be anywhere from 30 to 60 mph. It is also possible that, on an urban road, the speed limit may change several times over a short distance as the road progresses through a built-up area. This variation will have an effect on the skid resistance requirements of the asphalt to be chosen. Additionally, the traffic loading may vary considerably depending on the type of urban area the road is serving. Areas near industrial estates will have a completely different traffic profile to those which are serving housing estates. There is again a balance to be struck between traffic loading and the skid resistance of the asphalt which must be made on a road by road basis at a local level.

The properties generally required of urban roads are given in Table 5.2.

5.2.3 Rural Roads

These roads (Figure 5.3) generally do not generally carry high traffic volumes, although many of them will be busy during commuting times and

Table 5.1 Properties Required of Motorways and Strategic Dual Carriageways

Property	Base	Binder course	Surface course
Deformation resistance	Not relevant	Good	Good
Resistance to cracking	Moderate	Moderate	Moderate
Permeability	Impermeable	Impermeable	Impermeable or permeable
Water damage resistance	Moderate	Moderate	Good
Stiffness	Good	Moderate	Not relevant
Resistance to fatigue	Good	Good	Good
Tensile strength	Moderate	Moderate	Not relevant
Skid resistance	Not relevant	Not relevant	Good
Cohesion	Not relevant	Not relevant	Moderate
Tyre/road noise	Not relevant	Not relevant	Good
Spray generation	Not relevant	Not relevant	Good
Chemical damage resistance	Not relevant	Not relevant	Moderate
Adhesion*	Moderate	Good	Good
Durability	Good	Good	Good
Funding	Not generally restricted		

* With layers above and/or below

be quiet during the rest of the time. They will usually only carry a limited amount of heavy goods vehicles; however, in some locations they will be regularly used by farming equipment. They consist of a combination of both reasonably wide two lanes down to single lane with passing passes and every width in between. This changing often means that the wheel track is not in what would normally be the conventional place, often intensifying any loading down a centre line which is usually where any longitudinal construction joint is placed. Many of these roads will have been widened over the years by the addition of small strips at either side which will again pose further issues to be considered. These narrow strips often sink over the years due to their construction method, which will induce a longitudinal crack as well as undulations along the length of the carriageway, both of which need special consideration.

The properties generally required of rural roads are given in Table 5.3.

Figure 5.2 Typical urban road.

5.2.4 Residential Roads

Residential roads (Figure 5.4) tend to be more straight-forward in terms of traffic loading and speed. They are mostly designed to take cars with the occasional heavy goods vehicle. The speed limit on these roads is generally 30 mph but there are many of these roads with a 20-mph speed limit.

The properties generally required of residential roads are given in Table 5.4.

5.2.5 Industrial Estate Roads

Industrial estate roads are also relatively straight-forward in that the bulk of the traffic they take is heavy goods vehicles at relatively slow speeds. Most of the industrial estate roads will be subject to a speed limit of less than 40 mph; however, they will be subject to slow moving heavy traffic that is likely to be turning in tight locations. This turning will have the effect of some of the wheels scrubbing across the asphalt in an abrasive manner, totally different to the usual action of tyres on asphalt.

The properties generally required of industrial estate roads are given in Table 5.5.

Table 5.2 Properties Required of Urban Roads

Property	Base	Binder course	Surface course
Deformation resistance	Not relevant	Good	Good
Resistance to cracking	Moderate	Moderate	Moderate
Permeability	Impermeable unless SUDS	Impermeable unless SUDS	Impermeable or permeable
Water damage resistance	Moderate	Moderate	Good
Stiffness	Good	Moderate	Not relevant
Resistance to fatigue	Good	Good	Good
Tensile strength	Moderate	Moderate	Not relevant
Skid resistance	Not relevant	Not relevant	Good
Cohesion	Not relevant	Not relevant	Moderate
Tyre/road noise	Not relevant	Not relevant	Good
Spray generation	Not relevant	Not relevant	Good
Chemical damage resistance	Not relevant	Not relevant	Moderate
Adhesion	Moderate	Good	Good
Durability	Good	Good	Good
Funding	Often restricted		

SUDS, sustainable drainage systems.

5.2.6 Bridge Decks

Bridge decks can be within many of the above-mentioned roads and can be of a relatively short length to many miles in length. They fall into two categories, being either steel decks or concrete slab decks, both of which react in different ways to asphalt. It is usual for both of these structures to apply a waterproofing layer immediately to the bridge deck before the asphalt layers are applied. This layer is to prevent any water which may drain through or around the asphalt from coming into contact with the bridge deck and causing damage to the deck itself. There are numerous waterproofing systems in use so the choice of the first asphalt layer has to take this into account because this system will be the layer it bonds to. The thickness of the asphalt layers on bridge decks is also kept to a minimum – weight on an elevated bridge deck is crucial and the more asphalt (and therefore the more weight), the stronger, and more expensive the structure must be to take this load. Therefore, bridge decks are usually

Figure 5.3 Typical rural road.

Table 5.3 Properties Required of Rural Roads

Property	Base	Binder course	Surface course
Deformation resistance	Not relevant	Moderate	Moderate
Resistance to cracking	Moderate	Moderate	Moderate
Permeability	Impermeable	Impermeable	Impermeable
Water damage resistance	Moderate	Moderate	Good
Stiffness	Good	Moderate	Not relevant
Resistance to fatigue	Good	Good	Good
Tensile strength	Moderate	Moderate	Not relevant
Skid resistance	Not relevant	Not relevant	Good
Cohesion	Not relevant	Not relevant	Moderate
Tyre/road noise	Not relevant	Not relevant	Moderate
Spray generation	Not relevant	Not relevant	Moderate
Chemical damage resistance	Not relevant	Not relevant	Moderate
Adhesion	Moderate	Good	Good
Durability	Good	Good	Good
Funding	Generally restricted		

Figure 5.4 Typical residential road (with pothole).

Table 5.4 Properties Required of Residential Roads

Property	Base	Binder course	Surface course
Deformation resistance	Not relevant	Moderate	Moderate
Resistance to cracking	Moderate	Moderate	Moderate
Permeability	Impermeable	Impermeable	Impermeable
Water damage resistance	Moderate	Moderate	Moderate
Stiffness	Moderate	Moderate	Not relevant
Resistance to fatigue	Moderate	Moderate	Moderate
Tensile strength	Moderate	Moderate	Not relevant
Skid resistance	Not relevant	Not relevant	Moderate
Cohesion	Not relevant	Not relevant	Low
Tyre/road noise	Not relevant	Not relevant	Moderate
Spray generation	Not relevant	Not relevant	Low
Chemical damage resistance	Not relevant	Not relevant	Low
Adhesion	Moderate	Moderate	Moderate
Durability	Good	Good	Good
Funding	Often restricted		

limited to two layers and often in the case of steel decks just a single layer. The asphalt material characteristics for laying on these two types of decks also needs to be different in that the steel deck will be more flexible than a rigid concrete deck and also will have a higher rate of expansion and contraction due to climatic conditions.

The properties generally required of long bridge decks are given in Table 5.6.

Table 5.5 Properties Required of Industrial Estate Roads and Dock Trafficked Areas

Property	Base	Binder course	Surface course
Deformation resistance	Not relevant	Good	Good
Resistance to cracking	Moderate	Moderate	Moderate
Permeability	Impermeable	Impermeable	Impermeable
Water damage resistance	Moderate	Moderate	Good
Stiffness	Good	Moderate	Not relevant
Resistance to fatigue	Good	Good	Good
Tensile strength	Moderate	Moderate	Not relevant
Skid resistance	Not relevant	Not relevant	Moderate
Cohesion	Not relevant	Not relevant	Good
Tyre/road noise	Not relevant	Not relevant	Moderate
Spray generation	Not relevant	Not relevant	Moderate
Chemical damage resistance	Not relevant	Not relevant	Good
Adhesion	Moderate	Good	Good
Durability	Good	Good	Good
Funding	Can be restricted		

Table 5.6 Properties Required of Long Bridge Decks

Property	Base	Binder course	Surface course
Deformation resistance	Not relevant	Good	Good
Resistance to cracking	Moderate	Moderate	Moderate
Permeability	Impermeable	Impermeable	Impermeable
Water damage resistance	Moderate	Moderate	Good
Stiffness	Good	Moderate	Not relevant
Resistance to fatigue	Good	Good	Good
Tensile strength	Moderate	Moderate	Not relevant
Skid resistance	Not relevant	Not relevant	Good
Cohesion	Not relevant	Not relevant	Moderate
Tyre/road noise	Not relevant	Not relevant	Good
Spray generation	Not relevant	Not relevant	Good
Chemical damage resistance	Not relevant	Not relevant	Moderate
Adhesion	Good	Good	Good
Durability	Good	Good	Good
Funding	Not generally restricted		

5.2.7 Tunnels

Tunnels give specific issues for asphalt surfacing in that the work is all under-taken in a confined space and closure is usually for a short time period. The traffic in tunnels is heavily canalised because the vehicles are not generally allowed to overtake. In addition to the channelisation, they are often rela-tively slow-moving with short distances between them. The ambient temper-atures in a tunnel are generally higher than outside and are also less prone to changes. This higher temperature means that the asphalt to be used will likely be trafficked consistently at a higher temperature than normal over a longer period of time with a shorter relax time between loadings.

The properties generally required in tunnels are given in Table 5.7.

5.3 Airfields

5.3.1 Runways

Runways have two distinctly differing stresses, the first being where the planes land and then the rest of the runway is subject to moving traffic which is either slowing down or speeding up. The landing area is relatively small in terms of the overall runway and it is usual to pave the whole area with the

Table 5.7 Properties Required in Tunnels

Property	Base	Binder course	Surface course
Deformation resistance	Not relevant	Good	Good
Resistance to cracking	Moderate	Moderate	Moderate
Permeability	Impermeable	Impermeable	Impermeable
Water damage resistance	Moderate	Moderate	Good
Stiffness	Moderate	Moderate	Not relevant
Resistance to fatigue	Good	Good	Good
Tensile strength	Moderate	Moderate	Not relevant
Skid resistance	Not relevant	Not relevant	Good
Cohesion	Not relevant	Not relevant	Moderate
Tyre/road noise	Not relevant	Not relevant	Good
Spray generation	Not relevant	Not relevant	Good
Chemical damage resistance	Not relevant	Not relevant	Moderate
Adhesion	Good	Good	Good
Durability	Good	Good	Good
Funding	Not generally restricted		

same material. The tyre pressure and loading on aircraft are significantly higher than normal vehicle traffic on the normal highways although there is always a gap between loadings, which is not the case for many highways.

The properties generally required of airfield runways are given in Table 5.8.

5.3.2 Taxiways

Taxiways (Figure 5.5) have significantly more loading than a runway in that the aircraft are always slow moving. In addition to this difference, they are often subject to stationary aircraft, especially close to the runway when the aircraft queue for take-off. However, they are not subject to impact loadings which are present in runways. The effect of slow-moving heavy loads is close to that of industrial areas but with significantly higher loadings.

The properties generally required of airfield taxiways are given in Table 5.8.

5.3.3 Helipads

The main stresses in helipads are static and impact loading. There is very little, if any, moving traffic so many of the considerations normally

Table 5.8 Properties Required of Airfield Runways and Taxiways

Property	Base	Binder course	Surface course
Deformation resistance	Not relevant	Good	Good
Resistance to cracking	Moderate	Moderate	Moderate
Permeability	Impermeable	Impermeable	Impermeable
Water damage resistance	Moderate	Moderate	Good
Stiffness	Good	Moderate	Not relevant
Resistance to fatigue	Good	Good	Good
Tensile strength	Moderate	Moderate	Not relevant
Skid resistance	Not relevant	Not relevant	Good
Cohesion	Not relevant	Not relevant	Good
Tyre/road noise	Not relevant	Not relevant	Moderate
Spray generation	Not relevant	Not relevant	Good
Chemical damage resistance	Not relevant	Not relevant	Moderate
Adhesion	Moderate	Good	Good
Durability	Good	Good	Good
Funding	Not generally restricted		

Figure 5.5 Military taxiway.

associated with the use of asphalt materials do not apply. The only real consideration in specifying asphalt for a helipad is point and static loading. This loading will likely be concentrated in a very small area due to the helicopters all landing in the middle of the pad. Most pads will not be subject to excessively heavy loadings, but it will be repeated in the same small area and will be of a continuous nature while the helicopter is parked.

The properties generally required of helipads are given in Table 5.9.

5.3.4 Refuelling Areas

The threat of fuel spillage adds a further complication to the choice of asphalt material selection. Not only must the asphalt be able to withstand the stresses involved with heavy static loadings of aircraft and vehicles, but it must also be able to withstand the action of spilt fuel. The usual binder in asphalt is petroleum-based and, therefore, when it comes into contact with fuel the two will mix and the binder in the asphalt will become softer and likely to fail quicker. Specially, modified binders should be selected in these locations.

The properties generally required of airfield refuelling areas are given in Table 5.10.

Table 5.9 Properties Required of Helipads

Property	Base	Binder course	Surface course
Deformation resistance	Not relevant	Good	Good
Resistance to cracking	Moderate	Moderate	Moderate
Permeability	Impermeable	Impermeable	Impermeable
Water damage resistance	Moderate	Moderate	Good
Stiffness	Good	Moderate	Not relevant
Resistance to fatigue	Good	Good	Good
Tensile strength	Moderate	Moderate	Not relevant
Skid resistance	Not relevant	Not relevant	Moderate
Cohesion	Not relevant	Not relevant	Good
Tyre/road noise	Not relevant	Not relevant	Not relevant
Spray generation	Not relevant	Not relevant	Not relevant
Chemical damage resistance	Not relevant	Not relevant	Moderate
Adhesion	Moderate	Good	Good
Durability	Good	Good	Good
Funding	Not generally restricted		

Table 5.10 Properties Required of Refuelling Areas

Property	Base	Binder course	Surface course
Deformation resistance	Not relevant	Good	Good
Resistance to cracking	Moderate	Moderate	Moderate
Permeability	Impermeable	Impermeable	Impermeable
Water damage resistance	Moderate	Moderate	Good
Stiffness	Good	Moderate	Not relevant
Resistance to fatigue	Good	Good	Good
Tensile strength	Moderate	Moderate	Not relevant
Skid resistance	Not relevant	Not relevant	Moderate
Cohesion	Not relevant	Not relevant	Moderate
Tyre/road noise	Not relevant	Not relevant	Moderate
Spray generation	Not relevant	Not relevant	Moderate
Chemical damage resistance	Not relevant	Not relevant	Good
Adhesion	Good	Good	Good
Durability	Good	Good	Good
Funding	Not generally restricted		

5.4 Footways and Cycleways

5.4.1 Footways

Footway material does not generally take much in the way of loading because it is usually restricted to pedestrian traffic which, in terms of loading, is very small. Therefore, the characteristic that plays the greatest role in asphalt material selection for footways is usually one of aesthetics rather than performance. This characteristic usually ensures that the surface course on most footways is of a small nominal size such as 3 mm or 6 mm. This reduced size also ties in with a reduction in the thickness of the layers which result from the low traffic loading. One point that should be considered, especially in the selection of the binder grade to be used, is the effect of high heel shoes and the point loading that they produce on the asphalt. The loading on the asphalt will be very high which, especially in very hot weather conditions, can lead to penetration of the heel into the asphalt surfacing.

Some footways, particularly urban footways, are regularly overrun by cars or even commercial vehicles and, therefore, have to be designed for this higher loading.

The properties generally required of footways are given in Table 5.11.

5.4.2 Cycleways

Cycleways are very similar to footways in the terms of traffic loading and, therefore, are also generally constructed with a small nominal size aggregate surface course. In fact, in many locations footways and cycleways are constructed together and often used as one entity. The main difference is that cycleways are usually constructed with a coloured asphalt surface course to differentiate them from the footways and/or traffic lanes. This colouration is achieved by the addition of a pigment to the bitumen and possibly the use of a coloured aggregate. Red is the general choice, but green is also sometimes used. The choice of the aggregate should be carefully considered because, when the coloured bitumen is worn away from the surface of the cycleway, the aggregate will show through. If a red or green aggregate is used with the appropriate pigment, this loss of colouration will be less of an issue but, when a conventional dark grey aggregate is used, the cycleway will lose most of its coloured effect.

The properties generally required of cycleways are given in Table 5.11.

5.5 Parking Areas

5.5.1 Multi-Storey Carparks

The asphalt laid in multi-storey carparks is generally only laid as one layer over the concrete constructed deck due to issues with overall height and

Table 5.11 Properties Required of Footways, Cycleways, Playgrounds and Tennis Courts

Property	Not overrun		Regularly overrun	
	Base	Surfacing	Base	Surfacing
Deformation resistance	Not relevant	Low	Not relevant	Moderate
Resistance to cracking	Moderate	Moderate	Moderate	Moderate
Permeability	Impermeable if not SUDS	Impermeable if not SUDS	Impermeable if not SUDS	Impermeable if not SUDS
Water damage resistance	Moderate	Good	Moderate	Good
Stiffness	Low	Not relevant	Moderate	Not relevant
Resistance to fatigue	Moderate	Moderate	Moderate	Moderate
Tensile strength	Moderate	Not relevant	Moderate	Not relevant
Skid resistance	Not relevant	Moderate	Not relevant	Moderate
Cohesion	Not relevant	Low	Not relevant	Moderate
Tyre/road noise	Not relevant	Low	Not relevant	Low
Spray generation	Not relevant	Low	Not relevant	Low
Chemical damage resistance	Not relevant	Low	Not relevant	Low
Adhesion	Moderate	Moderate	Moderate	Good
Durability	Moderate	Moderate	Moderate	Moderate
Funding	Often restricted		Can be restricted	

SUDS, sustainable drainage systems.

also the weight. The traffic in these carparks is channelised on the main throughways which then also need to take the stress of slow turning traffic into the parking bays. The parking bays themselves are also subject to the same stresses but, additionally, have the point loading of vehicles being parked, sometimes for long periods. This loading means that the choice of asphalt is governed by the need to withstand scuffing and wheel tracking. These requirements will usually result in the choice of a modified binder to ensure the performance needed from a thin surface layer.

The properties generally required of multi-storey carparks are given in the final column of Table 5.12.

Table 5.12 Properties Required of Carparks

Property	Base	Binder course	Surface course
Deformation resistance	Not relevant	Moderate	Moderate
Resistance to cracking	Moderate	Moderate	Moderate
Permeability	Impermeable unless SUDS	Impermeable unless SUDS	Impermeable unless SUDS
Water damage resistance	Moderate	Moderate	Good
Stiffness	Moderate	Moderate	Not relevant
Resistance to fatigue	Moderate	Moderate	Moderate
Tensile strength	Moderate	Moderate	Not relevant
Skid resistance	Not relevant	Not relevant	Moderate
Cohesion	Not relevant	Not relevant	Moderate
Tyre/road noise	Not relevant	Not relevant	Low
Spray generation	Not relevant	Not relevant	Low
Chemical damage resistance	Not relevant	Not relevant	Moderate
Adhesion	Moderate	Good	Good
Durability	Good	Good	Good
Funding	Can be restricted		

SUDS, sustainable drainage systems.

5.5.2 External Carparks

External carparks are subject to the same stresses as multi-storey carparks but are constructed in a conventional way with more and thicker asphalt layers. While the same requirements are needed in the surface course, the requirements of the lower layers need the same properties as those in residential or urban roads.

The properties generally required of external carparks are given in Table 5.12.

5.5.3 Lorry Parks

The same considerations apply for lorry parks as in external carparks with the added complications of the additional heavy loading and often fuel or oil spillage. The former is considered in the design thickness calculations while the latter needs to be addressed in the material choice. Therefore, it is usually advisable to use a specially modified binder which will be able to cope with the high loading and turning stresses but also give some resistance to fuel and oil spillage. Grouted macadams are often considered for these applications.

The properties generally required of lorry parks are given in Table 5.13.

5.5.4 Sustainable Drainage Schemes

Sustainable drainage systems, or SuDS as they are commonly called, are asphalt areas which are designed to allow the water to enter the asphalt during heavy rain and then to dissipate slowly to help prevent flooding. Therefore, these materials need to be carefully designed to be as open as possible, which then gives problems in terms of performance under traffic.

The properties generally required of sustainable drainage schemes are given in Table 5.14.

5.5.5 Fuel Stations

As with aircraft refuelling areas, the threat of fuel spillage adds a further complication to the choice of asphalt material selection. The asphalt must be able to withstand the stresses involved with static loading and turning

Table 5.13 Properties Required of Lorry Parks

Property	Base	Binder course	Surface course
Deformation resistance	Not relevant	Good	Good
Resistance to cracking	Moderate	Moderate	Moderate
Permeability	Impermeable	Impermeable	Impermeable
Water damage resistance	Moderate	Moderate	Good
Stiffness	Moderate	Moderate	Not relevant
Resistance to fatigue	Moderate	Moderate	Moderate
Tensile strength	Moderate	Moderate	Not relevant
Skid resistance	Not relevant	Not relevant	Moderate
Cohesion	Not relevant	Not relevant	Good
Tyre/road noise	Not relevant	Not relevant	Low
Spray generation	Not relevant	Not relevant	Low
Chemical damage resistance	Not relevant	Not relevant	Moderate
Adhesion	Moderate	Good	Good
Durability	Good	Good	Good
Funding	Can be restricted		

Table 5.14 Properties Required of Sustainable Drainage Schemes

Property	Base	Binder course	Surface course
Deformation resistance	Not relevant	Moderate	Moderate
Resistance to cracking	Moderate	Moderate	Moderate
Permeability	Permeable	Permeable	Permeable
Water damage resistance	Good	Good	Good
Stiffness	Moderate	Moderate	Not relevant
Resistance to fatigue	Moderate	Moderate	Moderate
Tensile strength	Moderate	Moderate	Not relevant
Skid resistance	Not relevant	Not relevant	Moderate
Cohesion	Not relevant	Not relevant	Moderate
Tyre/road noise	Not relevant	Not relevant	Low
Spray generation	Not relevant	Not relevant	Moderate
Chemical damage resistance	Moderate	Moderate	Moderate
Adhesion	Good	Good	Good
Durability	Good	Good	Good
Funding	Can be restricted		

of cars and heavy goods vehicles and it must also be able to withstand the action of spilt fuel. The usual binder in asphalt is petroleum-based and, therefore, when it comes into contact with fuel, the two will mix and the binder in the asphalt will become softer and likely to fail quicker. Specially modified binders should be selected in these locations.

The properties generally required of fuel stations are given in Table 5.10.

5.6 Docks

5.6.1 Storage Areas

These areas (Figure 5.6) are subject to action of heavy-duty fork lift trucks together with the point loading of containers standing on the surface. These fork lift trucks have small wheels which subject the asphalt to extreme pressures. When the containers are stored on the area, they are supported at the four corner points so that the point loading in these areas is very high, especially if several containers are stacked on top of each other. These areas are likely to be subjected to the highest stresses of any paved area.

The properties generally required of dock storage areas are given in Table 5.15.

Figure 5.6 Example of storage area of docks.

Table 5.15 Properties Required of Dock Storage Areas

Property	Base	Binder course	Surface course
Deformation resistance	Not relevant	Good	Good
Resistance to cracking	Moderate	Moderate	Moderate
Permeability	Impermeable	Impermeable	Impermeable
Water damage resistance	Moderate	Good	Good
Stiffness	Moderate	Moderate	Not relevant
Resistance to fatigue	Good	Good	Good
Tensile strength	Moderate	Moderate	Not relevant
Skid resistance	Not relevant	Not relevant	Moderate
Cohesion	Not relevant	Not relevant	Good
Tyre/road noise	Not relevant	Not relevant	Low
Spray generation	Not relevant	Not relevant	Moderate
Chemical damage resistance	Moderate	Moderate	Good
Adhesion	Moderate	Good	Good
Durability	Good	Good	Good
Funding	Not generally restricted		

5.6.2 Trafficked Lanes

These areas are relatively straight-forward in that the traffic they take is heavy goods vehicles at relatively slow speeds; however, they will also be subject to trafficking by high-capacity fork lift trucks. These have small wheels which subject the asphalt to very high stresses. Most of these roads will be subject to a speed limit of less than 30 mph; however, they will be subject to slow-moving heavy traffic that is likely to be turning in tight locations. This turning will have the effect of some of the wheels scrubbing across the asphalt in an abrasive manner, totally different to the usual action of tyres on asphalt.

The properties generally required of dock trafficked areas are given in Table 5.5.

5.7 Ancillary Purposes

5.7.1 Playgrounds

Playgrounds do not generally take much in the way of loading, but they can be subject to a scuffing action because they are often used for sports and have children playing on them. The installation of playground equipment on which children can climb and, therefore, fall off will pose a further issue because conventional asphalt is hard. Therefore, the characteristics that play the greatest role in asphalt material selection for footways are aesthetics and safety from falling rather than performance. These properties usually ensure that the surface course on most playgrounds is of a small nominal size such as 3 mm or 6 mm. This reduced size also ties in with a reduction in the thickness of the layers which result from the low traffic loading. The safety aspect must be considered because the surface must not be slippery, especially in the wet. The area around equipment should consider the removal of the aggregate and its from the mixture replacement by rubber pieces or some equivalent.

Some playgrounds are regularly overrun by cars or even commercial vehicles and, therefore, have to be designed for this higher loading.

The properties generally required of playground surface courses are given in Table 5.11.

5.7.2 Tennis Courts

Tennis and other sports courts are subject to harsh turning forces by the action of the feet of competitors turning sharply on the surface. The asphalt also needs to be as smooth as possible so that there is a consistent surface on which the ball can bounce. Because these courts need to be useable in many weather conditions and quickly after rain as well as needing to be flat, an asphalt which will allow the water to drain through quickly needs to be selected. For aesthetic purposes coloured asphalts are often used so care needs to be taken with the colour of the aggregate used. It is often felt necessary to use some form of modified bitumen in order to give the performance requirements needed to stand up to the heavy scuffing action.

The properties generally required of tennis court surface courses are given in Table 5.11.

5.7.3 Race Circuits

The asphalt used for the lower layers on race circuits (Figure 5.7) are those used for an urban road. The surface course is, however, a very specialised product and will vary depending on what vehicles the race circuit is designed for. Racing circuits these days can designed for everything from light go karts through motor bikes and formula cars to heavy trucks. The actions produced by the tyres on these vehicles all affect the surface in a similar way in that they have very little tread. This smoothness, together with the severe forces in braking and cornering at high speed, generate high shear forces in the material. In addition to this loading, the laying tolerance on most race circuits is very tight, especially formula tracks, so that the nominal size of the aggregate used should be as small as possible while still withstanding the high shear forces. The binder used in most circuits would be specified as a highly modified one in order to help combat these shear forces.

The properties generally required of race circuits are given in Table 5.16.

5.7.4 Tank Farm Bases

The asphalt material to be used for tank farms is different in concept to most other asphalt materials in that it is designed in order to be sufficiently soft to allow the tanks placed on it to bed down into the asphalt. It is also designed without the use of any coarse aggregate in order to aid the bedding process, to prevent any chance of the aggregate damaging the tank and to prevent any high spots in the asphalt which may cause bridging of the tank when it is placed.

The properties generally required of tank farm bases are given in Table 5.17.

5.7.5 Dam and Canal Liners

The main criterion for these applications is a high resistance to water impregnation because they are designed to help keep the water in. This

Figure 5.7 Brooklands race track.

Table 5.16 Properties Required of Race Circuits

Property	Base	Binder course	Surface course
Deformation resistance	Not relevant	Good	Good
Resistance to cracking	Good	Good	Good
Permeability	Impermeable	Impermeable	Impermeable
Water damage resistance	Moderate	Good	Good
Stiffness	Moderate	Moderate	Not relevant
Resistance to fatigue	Good	Good	Good
Tensile strength	Moderate	Moderate	Not relevant
Skid resistance	Not relevant	Not relevant	Good
Cohesion	Not relevant	Not relevant	Good
Tyre/road noise	Not relevant	Not relevant	Moderate
Spray generation	Not relevant	Not relevant	Good
Chemical damage resistance	Moderate	Moderate	Good
Adhesion	Moderate	Good	Good
Durability	Good	Good	Good
Funding	Not generally restricted		

Table 5.17 Properties Required of Tank Farm Bases and Dam and Canal Liners

Property	Tank farm bases	Dam and canal liners
Deformation resistance	Moderate	Not relevant
Resistance to cracking	Good	Good
Permeability	Impermeable	Impermeable
Water damage resistance	Good	Good
Stiffness	Not relevant	Not relevant
Resistance to fatigue	Good	Good
Tensile strength	Not relevant	Not relevant
Skid resistance	Not relevant	Not relevant
Cohesion	Not relevant	Not relevant
Tyre/road noise	Not relevant	Not relevant
Spray generation	Not relevant	Not relevant
Chemical damage resistance	Good	Good
Adhesion	Not relevant	Not relevant
Durability	Good	Good
Funding	Can be restricted	Not generally restricted

requirement needs a material with virtually no air voids in the mixture. It also needs a mixture that will not be damaged by water. The mixtures used for these applications are, therefore, generally a blend of a fine aggregate (less than 3 mm) with a high content of bitumen. The high bitumen content improves the water-proofing ability and reduces the air voids content.

The properties generally required of dam and canal liners are given in Table 5.17.

5.7.6 Coastal Erosion

Asphalt is used in protection against coastal erosion, but with a range of mixtures with titles that are different to those used for other applications. These mixture titles include open stone asphalt, lean sand asphalt and grouted stone. These asphalt mixtures are outside the scope of this book; readers will need to seek advice from an asphalt specialist in hydraulic engineering for advice on mixture selection.

Chapter 6

External Influences

6.1 Site

6.1.1 Substrate

Each layer in a pavement can only be as durable as the layers below it. If a layer below is weak, it will require the layer itself to be stiffer or thicker to compensate provided that extra stiffness does not make it too brittle. The ultimate case is the substrate on which the pavement rests and that substrate can vary from solid rock to peat or loose sand. If the substrate is considered to be too weak, it will be dug out and replaced by a firmer foundation that will take the expected loads.

In terms of selecting material, the primary case is use of EME2 (Sub-Section 4.2.5) as the base or binder course which requires a stiff substrate because the stiffness of the material is provided at the expense of being relatively brittle.

6.1.2 Road Geometry

The road geometry, particularly roundabouts (Figure 6.1) and other forms of intersection, can affect the stresses applied to the pavement, particularly the surface, by affecting the speed at which vehicles can travel and the amount that they have to change direction. The surface course at intersections needs to have better resistance to skidding and scuffing than less stressed areas. These properties can be provided by surface treatments, in particular resin-based high-friction surfacing systems, or by choice of asphalt type and aggregate properties. Generally, the aggregate properties required are higher polished stone value (PSV) for skid resistance and smaller maximum size for scuffing resistance.

The geometry is not generally a feature for other paved areas.

The geometry will exaggerate the need for properties in the surfacing as follows:

Figure 6.1 Roundabout with scuffing loss.

- Deformation resistance – Marginally increased because of traffic travelling more slowly.
- Resistance to cracking – No change.
- Permeability – No change.
- Water damage resistance – No change.
- Stiffness – No change.
- Resistance to fatigue – No change.
- Tensile strength – No change.
- Skid resistance – Increased with more turning and braking.
- Cohesion – Increased with more turning and braking.
- Tyre/road noise – No change.
- Spray generation – No change.
- Chemical damage resistance – No change.
- Adhesion – Marginally increased.
- Durability – No change.
- Funding – No change.

6.1.3 Gradient

Ascending steep gradients will require vehicles, particularly heavily loaded lorries, to apply greater force, as well as restricting the speed, while descending steep gradients (Figure 6.2) will require the same vehicles to apply their brakes to remain under control. Therefore, the asphalt, particularly the surface course, will need to have good resistance to the shear force applied, the extent increasing with steepness of the gradient.

The gradient can also have an effect on the asphalt temperature if the slope faces towards the equator because the surface will be exposed more directly to the sun (Sub-Section 6.3.1). This heating will be more pronounced when the asphalt is relatively new (when it is more susceptible to deformation because the binder has not aged) and, therefore, blacker.

Gradient is not generally a feature for other paved areas, particularly airfields and docks.

The gradient will exaggerate the need for properties in the surfacing as follows:

Figure 6.2 Example of a trunk road with descending gradient.

- Deformation resistance – Marginally increased because of heavy traffic travelling more slowly uphill.
- Resistance to cracking – No change.
- Permeability – No change.
- Water damage resistance – No change.
- Stiffness – No change.
- Resistance to fatigue – No change.
- Tensile strength – No change.
- Skid resistance – No change.
- Cohesion – Marginally increased with more traction required.
- Tyre/road noise – No change.
- Spray generation – No change.
- Chemical damage resistance – No change.
- Adhesion – Marginally increased.
- Durability – No change.
- Funding – No change.

6.1.4 Bends

Bends in non-intersection lengths of roads will add stresses similarly to intersections (Sub-Section 6.1.2), if generally to a lesser extent because the radius is not often as tight. Also, the direction of bend does not always change over such short distances. Nevertheless, the surface course on sharp bends also needs to have better resistance to skidding and scuffing than less stressed areas, although rarely to the extent as to require high-friction surfacing systems.

- Deformation resistance – No change.
- Resistance to cracking – No change.
- Permeability – No change.
- Water damage resistance – No change.
- Stiffness – No change.
- Resistance to fatigue – Marginally increased.
- Tensile strength – No change.
- Skid resistance – Increased with more turning.
- Cohesion – Increased with more turning.
- Tyre/road noise – No change.
- Spray generation – No change.
- Chemical damage resistance – No change.
- Adhesion – Marginally increased.
- Durability – No change.
- Funding – No change.

6.2 Traffic

6.2.1 General

Traffic loading is a combination of the type of traffic, the volume of traffic, its regularity and the speed of the vehicles. The loading is generally compressive through the vehicle tyres but there will also be tensile and torsional stresses from turning vehicles or tracked vehicles.

6.2.2 Type of Traffic

The type of traffic that travels on pavements will affect the type of asphalt required to successfully produce that pavement.

Most roads carry mostly cars with a significant proportion of commercial vehicles. Deformation is regarded as being proportional to the fourth (or even higher) power of the wheel load (Cebon, 2007) so the higher the proportion of the heavy commercial vehicles, the more deformation-resistant the asphalt needs to be. Similarly, airfields carrying larger aircraft will require more deformation-resistant asphalt, but the load will be exacerbated by faster aircraft, particularly military aircraft, requiring narrower tyres to fit within the thin wings which have to have higher tyre pressures. For docks, the vehicles include a significant proportion of high-capacity fork lift trucks with small wheels which subject the asphalt to very high stresses (Sub-Section 5.6.1).

The presence of tracked vehicles on paved areas will impose very high shear stresses to the surface course, but there are not many sites which such vehicles regularly traverse. On such sites, considerable care is needed in order to produce a durable solution.

Those sites with little or no vehicular traffic, generally footpaths, cycleways and playgrounds that are not regularly overrun, will effectively have no traffic loading.

6.2.3 Volume and Regularity of Traffic

The volume of traffic is important but it needs to be combined with its regularity. It is the average volume that is most important. There are some advantages in rest periods with lower traffic when the asphalt can recover from the loading periods with heavier traffic. However, the rest period between vehicles is more beneficial than the rest periods between different times of the day or year.

A high volume of traffic will exaggerate the need for properties in the surfacing as follows:

- Deformation resistance – Increased because more traffic passing.
- Resistance to cracking – No change.
- Permeability – No change.
- Water damage resistance – No change.
- Stiffness – No change.
- Resistance to fatigue – Increased.
- Tensile strength – No change.
- Skid resistance – Increased with more traffic.
- Cohesion – Increased with more traffic.
- Tyre/road noise – Increased.
- Spray generation – Increased.
- Chemical damage resistance – No change.
- Adhesion – Increased.
- Durability – No change.
- Funding – More generally available.

6.2.4 *Speed of Vehicles*

The slower a vehicle is moving, the longer time its load is applied to each section of pavement. Therefore, slow-moving vehicles increase the loading and the heaviest vehicles tend to be the slowest. The speed of the traffic will depend on a number of facts including the road geometry, the speed limits and any maintenance works being undertaken at the time.

The speed of traffic will exaggerate the need for properties in the surfacing as follows:

- Deformation resistance – No change.
- Resistance to cracking – No change.
- Permeability – No change.
- Water damage resistance – No change.
- Stiffness – No change.
- Resistance to fatigue – No change.
- Tensile strength – No change.
- Skid resistance – Increased with greater consequences from lack of grip.
- Cohesion – Increased with greater effects from turning and braking.
- Tyre/road noise – Increased.
- Spray generation – Increased.
- Chemical damage resistance – No change.
- Adhesion – Marginally increased.
- Durability – No change.
- Funding – No change.

6.3 Environment

6.3.1 Ambient Temperature

The ambient temperature has an influence on the selection of asphalt mixture type in terms of its maximum, its minimum and its range. However, the average temperature does not generally vary very much within any region apart from minor changes with the exposure to the sun (Sub-Sections 6.1.3 and 6.3.3).

The maximum temperature affects the deformation resistance required of the surface course and, to a lesser extent, the binder course. The deformation of asphalt pavements occurs primarily during a relatively short period of the year when the temperatures are highest because, if noticeable deformation occurs at lower temperatures, the rutting will require treatment very quickly. Higher temperatures require the use of a stiffer binder and/or the use of stone-interlock rather than mastic mixture types.

The minimum temperature affects the flexibility and brittleness of the asphalt. As such, lower temperatures require the use of a softer binder and a binder rich (or mastic) asphalt mixture type. Because the requirements for high and low temperatures are contradictory, sites with a wide temperature range require a careful compromise.

High maximum temperatures will exaggerate the need for properties in the surfacing as follows:

- Deformation resistance – Increased because asphalt more deformation resistance.
- Resistance to cracking – Reduced because asphalt more fluid.
- Permeability – No change.
- Water damage resistance – No change.
- Stiffness – No change.
- Resistance to fatigue – Reduced because asphalt more fluid.
- Tensile strength – No change.
- Skid resistance – No change.
- Cohesion – no change.
- Tyre/road noise – No change.
- Spray generation – No change.
- Chemical damage resistance – No change.
- Adhesion – No change.
- Durability – No change.
- Funding – No change.

Low minimum temperatures will exaggerate the need for properties in the surfacing as follows:

- Deformation resistance – Decreased because asphalt more deformation-resistant.
- Resistance to cracking – Increased because asphalt more brittle.
- Permeability – No change.
- Water damage resistance – No change.
- Stiffness – No change.
- Resistance to fatigue – Increased because asphalt more brittle.
- Tensile strength – No change.
- Skid resistance – No change.
- Cohesion – No change.
- Tyre/road noise – No change.
- Spray generation – No change.
- Chemical damage resistance – No change.
- Adhesion – No change.
- Durability – No change.
- Funding – No change.

6.3.2 Damp Conditions

The most obvious locations for damp conditions are at fords, tidally covered access routes to islands or on flood plains which are allowed to flood. However, there are other locations, particularly in areas with a high water-table, where the pavement can be damp even without any precipitation and there are situations where the water is retained after precipitation. In such situations, the asphalt needs to be resistant to moisture damage which is generally enhanced by high binder contents and good aggregate/binder affinity.

Damp conditions will exaggerate the need for properties in the surfacing as follows:

- Deformation resistance – No change.
- Resistance to cracking – No change.
- Permeability – Permeable materials need to be more impermeable.
- Water damage resistance – Increased.
- Stiffness – No change.
- Resistance to fatigue – No change.
- Tensile strength – No change.
- Skid resistance – No change.
- Cohesion – No change.
- Tyre/road noise – No change.
- Spray generation – Marginally increased.
- Chemical damage resistance – No change.
- Adhesion – No change.

- Durability – No change.
- Funding – No change.

6.3.3 Shade

There are areas on many asphalt pavements which are shaded for part or all of the day (Figure 6.3). Such conditions tend to make the pavement cooler (Sub-Section 6.3.1) and damper (Sub-Section 6.3.2) than the surrounding areas.

Regular shade will exaggerate the need for properties in the surfacing as follows:

- Deformation resistance – No change.
- Resistance to cracking – No change.
- Permeability – No change.
- Water damage resistance – Marginally increased.
- Stiffness – No change.
- Resistance to fatigue – No change.
- Tensile strength – No change.

Figure 6.3 Example of a section of road regularly in shade.

- Skid resistance – Marginally increased.
- Cohesion – Marginally increased.
- Tyre/road noise – No change.
- Spray generation – No change.
- Chemical damage resistance – No change.
- Adhesion – Marginally increased.
- Durability – No change.
- Funding – No change.

6.3.4 Snow

Pavements covered by snow for extensive periods obviously tend to have low minimum temperatures (Sub-Section 6.3.1). However, studded tyres are often used in such areas which abrade the aggregate in the mixture. Stone mastic asphalt (Section 4.3) was developed in Germany in order to resist studded tyres which were subsequently banned in that country, although SMA is still used extensively there. Another mixture developed in Scandinavia was soft asphalt (Section 4.9) on a sacrificial basis because it can be laid in colder conditions.

Extended snow coverage will exaggerate the need for properties in the surfacing as follows:

- Deformation resistance – Decreased because asphalt more deformation-resistant.
- Resistance to cracking – Increased because asphalt more brittle.
- Permeability – No change.
- Water damage resistance – Increased because of thawing conditions.
- Stiffness – No change.
- Resistance to fatigue – Increased because asphalt more brittle.
- Tensile strength – No change.
- Skid resistance – No change.
- Cohesion – No change.
- Tyre/road noise – No change.
- Spray generation – No change.
- Chemical damage resistance – No change.
- Adhesion – Increased with snow clearing.
- Durability – No change.
- Funding – No change.

6.4 Economics

The factor that often is the most influential in the choice of the materials used in constructing a pavement is economics. Almost every pavement owner has insufficient funds to do all the construction and maintenance that it considers necessary to provide a network in perfect condition. Therefore, the pavement owner may have to sacrifice the long-term durability in order to provide an immediate solution. The aspects that can affect whether economics do play a part in the choice of asphalt type include the total funds available, the strategic importance of the particular pavement, the consequences of future failure and, least reputable of them, the political importance of the site.

References

Cebon, D (2007). Vehicle-generated road damage: A review. *Vehicle System Dynamics, International Journal of Vehicle Mechanics and Mobility*, 18(1-3), 107–150.

Chapter 7

Asphalt Mixture Selection

7.1 General Approach

In order to select an appropriate mixture for a particular situation, a series of questions about the site need to be considered. For the methodology proposed in this book, the questions are considered in the sequence shown in Figure 7.1.

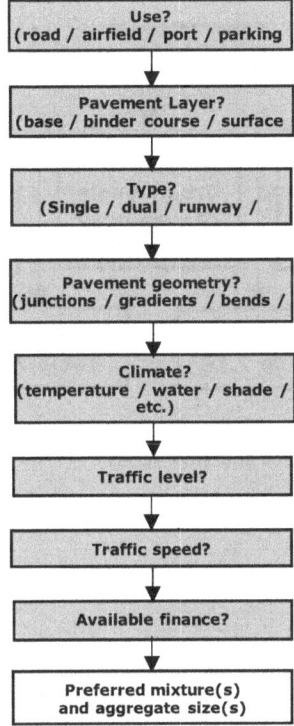

Figure 7.1 Basic flow-chart for select of preferred asphalt mixture(s).

The first two aspects are covered by the choice of section and sub-section, respectively. The remaining aspects are covered in a series of flow-charts leading to tables of suitable mixture types with several of those tables being referenced in more than one section or sub-section. Not all aspects are needed to be included in the selection for every situation.[1]

The choice found from the selection process is not necessarily a unique material for each set of conditions. Therefore, the authors have identified their preferred mixture type in **bold**.

7.2 Roads

7.2.1 Surface Course

A selection of suitable asphalt mixtures for use in the surface course of roads is provided by following the flow-charts starting with Figure 7.2 to identify the relevant table of mixture type for the situation.

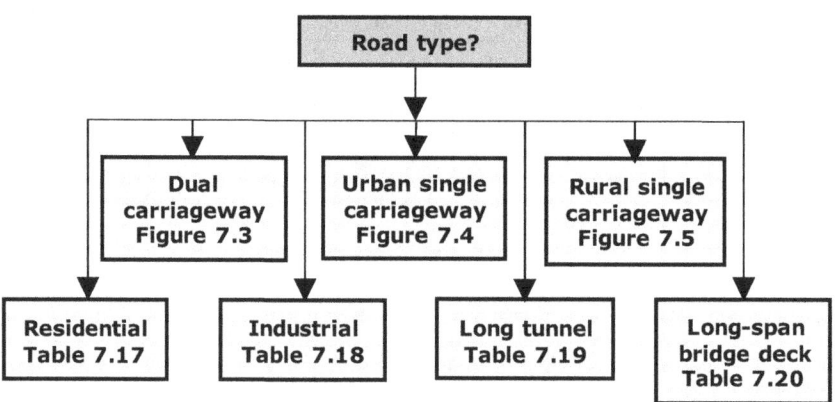

Figure 7.2 Flow-chart for surface course of roads by road type.

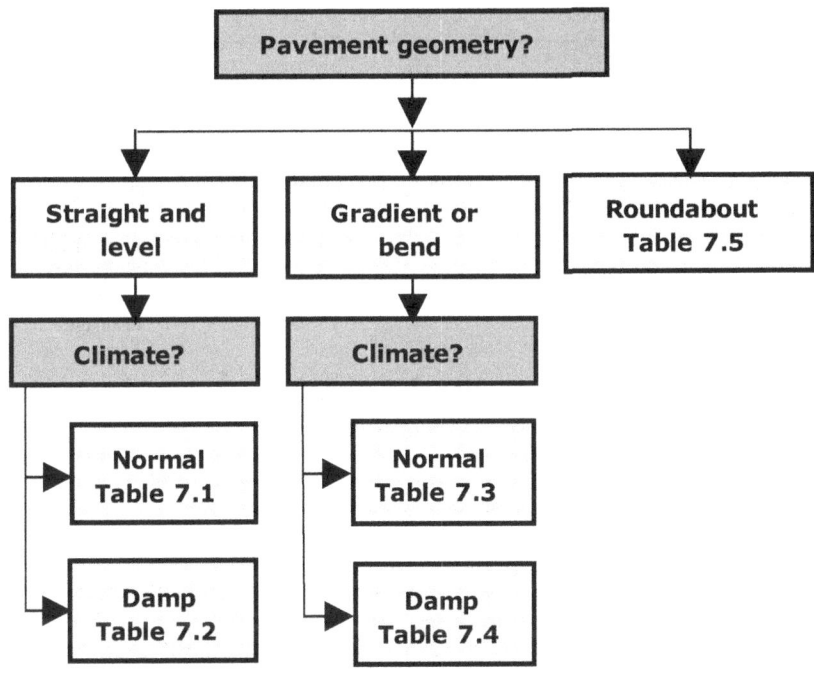

Figure 7.3 Flow-chart of ancillary purposes.

Table 7.1 Mixtures for Surface Course of Straight and Level Dual-Carriageway Roads

Traffic level	Traffic speed	Cost band	Mixture type	Suggested NMAS	Design specifics
Any	Any	Economy	AUTL		Not heavy traffic
		Standard	AC/MA	14 mm	–
			SMA		–
			BBTM		–
			PA	10 mm or 20 mm	Impermeable binder course and specific drainage
			PA/TL	10 mm & 20 mm	
		Premium	AC/SP		–
			HRA+PCC	14 mm	–

Table 7.2 Mixtures for Surface Course of Straight and Level Dual-Carriageway Roads in Damp Conditions

Traffic level	Traffic speed	Cost band	Mixture type	Suggested NMAS	Design specifics
Any	Any	Standard	AC/MA		—
			SMA	14 mm	—
			BBTM		—
			PA	10 mm or 20 mm	Impermeable binder course and specific drainage
			PA/TL	10 mm & 20 mm	
		Premium	AC/SP		—
			HRA+PCC	14 mm	—

Table 7.3 Mixtures for Surface Course of Dual-Carriageway Roads on Bends or Slopes

Traffic level	Traffic speed	Cost band	Mixture type	Suggested NMAS	Design specifics
Any	Any	Economy	AUTL		Not heavy traffic
		Standard	AC/MA	10 mm	—
			SMA		—
			BBTM		—
			PA	10 mm or 20 mm	Impermeable binder course and specific drainage
			PA/TL	10 mm & 20 mm	
		Premium	AC/SP		—
			HRA+PCC	10 mm	—

Table 7.4 Mixtures for Surface Course of Dual-Carriageway Roads on Bends or Slopes in Damp Conditions

Traffic level	Traffic speed	Cost band	Mixture type	Suggested NMAS	Design specifics
Any	Any	Standard	AC/MA		–
			SMA	10 mm	–
			BBTM		–
			PA	10 mm or 20 mm	Impermeable binder course and specific drainage
			PA/TL	10 mm & 20 mm	
		Premium	AC/SP		–
			HRA+PCC	10 mm	–

Table 7.5 Mixtures for Surface Course of Dual-Carriageway Roads and Single-Carriageway Urban Roads at Roundabouts Including in Damp Conditions and of Racing Circuits

Traffic level	Traffic speed	Cost band	Mixture type	Suggested NMAS	Design specifics
Any	Any	Economy	AC/MA		Reduced texture requirements
		Standard	SMA		
			BBTM	10 mm	
			HRA+PCC		
		Premium	AC/SP		

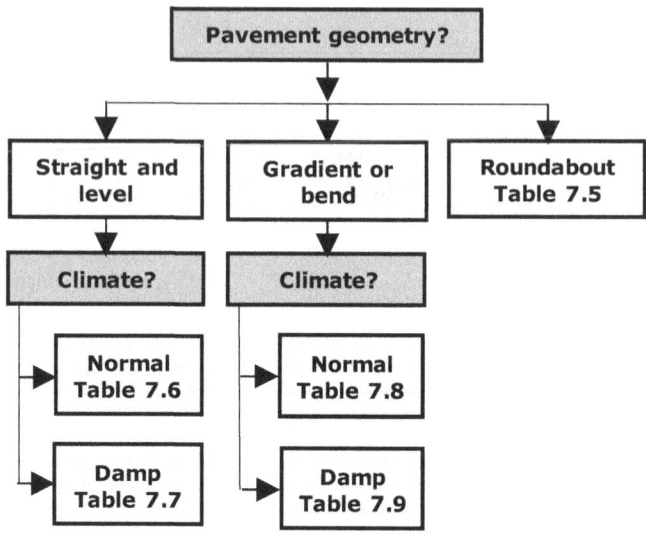

Figure 7.4 Flow-chart for surface course of urban single-carriageway roads.

Table 7.6 Mixtures for Surface Course of Straight and Level Urban Single-Carriageway Roads

Traffic level	Traffic speed	Cost band	Mixture type	Suggested NMAS	Design specifics
		Economy	AC/DBM		–
			AUTL		Not heavy traffic
			AC/MA	14 mm	–
		Standard	**SMA**		–
			BBTM		–
Any	Standard		PA	10 mm or 20 mm	Impermeable binder course
			PA/TL	10 mm & 20 mm	and specific drainage
		Premium	HRA+PCC		–
			AC/SP	14 mm	–

Table 7.7 Mixtures for Surface Course of Straight and Level Urban Single-Carriageway Roads in Damp Conditions

Traffic level	Traffic speed	Cost band	Mixture type	Suggested NMAS	Design specifics
		Economy	AC/DBM		–
			AC/MA	14 mm	–
		Standard	**SMA**		–
			BBTM		–
			PA	10 mm or 20 mm	Impermeable binder course
Any	Standard		PA/TL	10 mm & 20 mm	and specific drainage
		Premium	HRA+PCC		–
			AC/SP	14 mm	–

Table 7.8 Mixtures for Surface Course of Urban Single-Carriageway Roads on Bends or Slopes

Traffic level	Traffic speed	Cost band	Mixture type	Suggested NMAS	Design specifics
			AC/DBM		–
		Economy	AUTL		Not heavy traffic
			AC/MA	10 mm	–
		Standard	**SMA**		–
			BBTM		–
Any	Standard		PA	10 mm or 20 mm	Impermeable binder course
			PA/TL	10 mm & 20 mm	and specific drainage
		Premium	HRA+PCC		–
			AC/SP	10 mm	–

Table 7.9 Mixtures for Surface Course of Urban Single-Carriageway Roads on Bends or Slopes in Damp Conditions

Traffic level	Traffic speed	Cost band	Mixture type	Suggested NMAS	Design specifics
Any	Standard	Economy	AC/DBM		–
		Standard	AC/MA	10 mm	–
			SMA		–
			BBTM		–
			PA	10 mm or 20 mm	Impermeable binder course and specific drainage
			PA/TL	10 mm & 20 mm	
		Premium	HRA+PCC		–
			AC/SP	10 mm	–

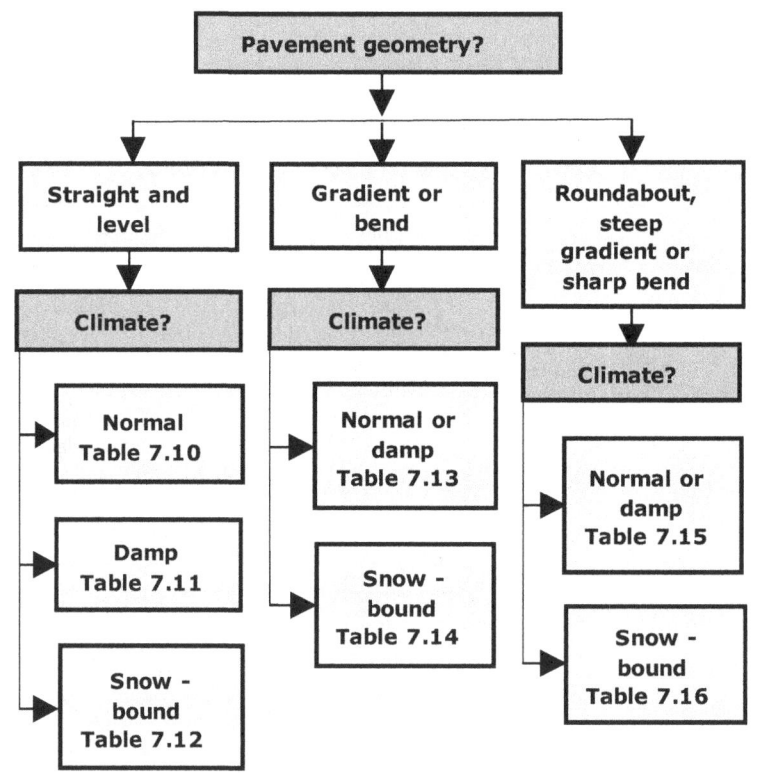

Figure 7.5 Flow-chart for surface course of rural single-carriageway roads.

Table 7.10 Mixtures for Surface Course of Straight and Level Rural Single-Carriageway Roads

Traffic level	Traffic speed	Cost band	Mixture type	Suggested NMAS	Design specifics
Heavy or medium	Any	Economy	AC/DBM		–
			AUTL		Not heavy traffic or slow traffic speed
		Standard	**SMA**	14 mm	–
			BBTM		–
		Premium	AC/MA		–
			HRA+PCC		–
Light	Slow	Economy	AC/DBM		–
		Standard	**SMA**		–
			BBTM		–
		Premium	AC/MA		–
			HRA+PCC		–
	Normal	Economy	AC/OBM		–
		Standard	**AC/DBM**	14 mm	–
			AUTL		–
		Premium	HRA		–
			AC/MA		–
			SMA		–
			BBTM		–
			HRA+PCC		–

Table 7.11 Mixtures for Surface Course of Straight and Level Rural Single-Carriageway Roads in Damp Conditions

Traffic level	Traffic speed	Cost band	Mixture type	Suggested NMAS	Design specifics
Heavy or medium	Any	Economy	AC/DBM		–
		Standard	**SMA**		–
			BBTM	14 mm	–
			HRA		–
		Premium	AC/MA		–
			HRA+PCC		–

(Continued)

Table 7.11 (Cont).

Traffic level	Traffic speed	Cost band	Mixture type	Suggested NMAS	Design specifics
Light	Slow	Economy	AC/DBM		–
			SMA		–
		Standard	BBTM		–
			HRA		–
		Premium	AC/MA		–
			HRA+PCC	14 mm	–
	Normal	Standard	**AC/DBM**		–
			HRA		–
		Premium	AC/MA		–
			SMA		–
			BBTM		–
			HRA+PCC		–

Table 7.12 Mixtures for Surface Course of Straight and Level Rural Single-Carriageway Roads in Snow-Bound Conditions

Traffic level	Traffic speed	Cost band	Mixture type	Suggested NMAS	Design specifics
Heavy or medium	Any	Economy	AC/DBM		–
			AUTL		Not heavy traffic or slow traffic speed
			SMA	14 mm	–
		Standard	BBTM		–
			HRA		–
		Premium	HRA+PCC		–
Light	Slow	Economy	AC/DBM		–
			SMA		–
		Standard	BBTM		–
			HRA		–
		Premium	HRA+PCC		–
	Normal	Standard	AC/DBM		–
			SA	14 mm	–
			HRA		–
			AUTL		–
		Premium	SMA		–
			BBTM		–
			HRA+PCC		–

Table 7.13 Mixtures for Surface Course of Rural Single-Carriageway Roads on Bends or Slopes Including in Damp Conditions

Traffic level	Traffic speed	Cost band	Mixture type	Suggested NMAS	Design specifics
Heavy or medium	Any	Economy	AC/DBM		–
		Standard	**SMA**		–
			BBTM	14 mm	–
			HRA		–
		Premium	AC/MA		–
			HRA+PCC		–
Light	Slow	Economy	AC/DBM		–
		Standard	**SMA**		–
			BBTM		–
			HRA		–
		Premium	AC/MA		–
			HRA+PCC		–
	Normal	Economy	AC/OBM	14 mm	–
		Standard	**AC/DBM**		–
			HRA		–
			AC/MA		–
			SMA		–
		Premium	BBTM		–
			HRA+PCC		–

Table 7.14 Mixtures for Surface Course of Rural Single-Carriageway Roads on Bends or Slopes in Snow-Bound Conditions

Traffic level	Traffic speed	Cost band	Mixture type	Suggested NMAS	Design specifics
Heavy or medium	Any	Economy	AC/DBM		–
			SA		–
		Standard	**SMA**		–
			BBTM	14 mm	–
			AUTL		–
			HRA		–
		Premium	HRA+PCC		–

(Continued)

Table 7.14 (Cont).

Traffic level	Traffic speed	Cost band	Mixture type	Suggested NMAS	Design specifics
Light	Slow	Economy	AC/DBM		–
			SA		–
		Standard	**SMA**		–
			BBTM		–
			AUTL		–
			HRA		–
		Premium	AC/MA		–
			HRA+PCC	14 mm	–
Light	Normal	Standard	AC/DBM		–
			SA		–
			HRA		–
		Premium	AC/MA		–
			SMA		–
			BBTM		–
			HRA+PCC		–

Table 7.15 Mixtures for Surface Course of Rural Single-Carriageway Roads on Round-abouts, Sharp Bends or Steep Slopes Including in Damp Conditions

Traffic level	Traffic speed	Cost band	Mixture type	Suggested NMAS	Design specifics
Any	Any	Standard	SMA		–
			BBTM		–
			HRA	10 mm	–
			AC/MA		–
		Premium	HRA+PCC		–

Table 7.16 Mixtures for Surface Course of Rural Single-Carriageway Roads on Round-abouts, Sharp Bends or Steep Slopes in Snow-Bound Conditions

Traffic level	Traffic speed	Cost band	Mixture type	Suggested NMAS	Design specifics
Heavy or medium	Any	Standard	SMA		–
			BBTM		–
			HRA		–
		Premium	AC/MA		–
			HRA+PCC		–
Light	Slow	Standard	SMA	10 mm	–
			BBTM		–
			HRA		–
		Premium	AC/MA		–
			HRA+PCC		–
	Normal	Standard	**SA**		–
		Premium	SMA		–
			BBTM		–
			HRA		–

Table 7.17 Mixtures for Surface Course of Residential Roads

Cost band	Mixture type	Suggested NMAS	Design specifics
Standard	**AC/DBM**		
	AC/MA		
	BBTM	6 mm or 10 mm	Limited texture requirements
Premium	SMA		
	HRA .		

Table 7.18 Mixtures for Surface Course of Industrial Roads and Trafficked Lanes of Docks

Traffic	Cost band	Mixture type	Suggested NMAS	Design specifics
High	Standard	AC/MA		–
		BBTM		–
		SMA		–
	Premium	AC/SP		–
		PA/GM		–
	Economy	AC/DBM		–
		AC/MA	10 mm or 14 mm	–
Medium or light	Standard	**BBTM**		–
		SMA		–
		AC/SP		–
	Premium	HRA+PCC		–
		PA/GM		–

Table 7.19 Mixtures for Surface Course of Long Tunnels

Cost band	Mixture type	Suggested NMAS	Design specifics
Standard	AC/MA		–
	BBTM		–
	SMA		–
	HRA+PCC	10 mm or 14 mm	–
	AC/SP		–
Premium	MA+PCC		–

Table 7.20 Mixtures for Surface Course of Long-Span Bridge Decks

Cost band	Mixture type	Suggested NMAS	Design specifics
Economy	AC/DBM		
	AUTL		
	AC/MA		
Standard	HRA+PCC	10 mm or less	Low air voids, limiting texture
	SMA		Requires separate waterproofing, more
	BBTM		important on more economical options
	AC/SP		
Premium	**MA+PCC**		

7.2.2 Binder Course

A selection of suitable asphalt mixtures for use in the binder course of roads is provided by following the flow-chart in Figure 7.6 to identify the relevant table of mixture type for the situation.

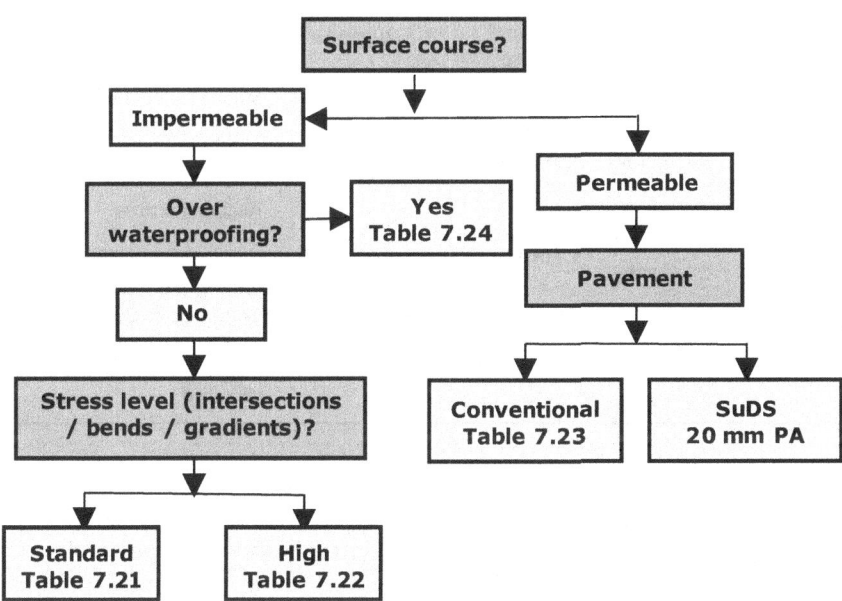

Figure 7.6 Flow-chart for binder course of roads.

Table 7.21 Mixtures for Binder Course of Low Stressed Roads with Conventional Surface Course and Race Circuits

Traffic	Cost band	Mixture type	Suggested NMAS	Design specifics
Heavy	Economy	AC/HDM or HMB		
	Standard	AC/MA SMA AC/SP	20 mm	Design mixture
	Premium	**AC/EME2**		Strong foundation needed
		HRA		Design mixture
Medium or light	Economy	**AC/DBM**	20 mm	
	Standard	AC/MA SMA		Recipe mixture
		HRA		
	Premium	PA		Only when groundwater cannot be excluded from pavement structure Appropriate drainage required

Table 7.22 Mixtures for Binder Course of High Stressed Roads with Conventional Surface Course, of Trafficked Lanes of Docks and of Motor Racing Circuits

Traffic	Cost band	Mixture type	Suggested NMAS	Design specifics
Heavy*	Economy	AC/HDM or HMB		
	Standard	AC/MA SMA AC/SP	20 mm	Design mixture
	Premium	**AC/EME2**		Strong foundation needed
		HRA		Design mixture
Medium or light	Economy	**AC/DBM**		
	Standard	AC/MA SMA AC/SP	20 mm	Design mixture
	Premium	HRA		

Table 7.23 Mixtures for Binder Course of Conventional Roads with Permeable Surface Course

Traffic	Cost band	Mixture type	Suggested NMAS	Design specifics
Heavy	Standard	AC/MA		Design mixture
		AC/SP		
	Premium	AC/EME2	20 mm	Strong foundation needed
		HRA		Design mixture
Medium or light	Standard	AC/MA		
		SMA		
		AC/SP	20 mm	Recipe mixture
	Premium	**HRA**		

Table 7.24 Mixtures for Layer Above a Waterproofing Layer of Roads over Bridges and Similar Structures

Traffic	Cost band	Mixture type	Suggested NMAS	Design specifics
All	Economy	HRA/SC *	4 mm	–
		AC/DBM		–
	Standard	AC/MA		–
		BBTM		–
		SMA	10 mm or less	–
		HRA		–
	Premium	MA		–

* Conventional but not recommended

7.2.3 Base

A selection of suitable asphalt mixtures for use in the base of roads is provided by the flow-chart in Figure 7.7 to identify the relevant table of mixture type for the situation.

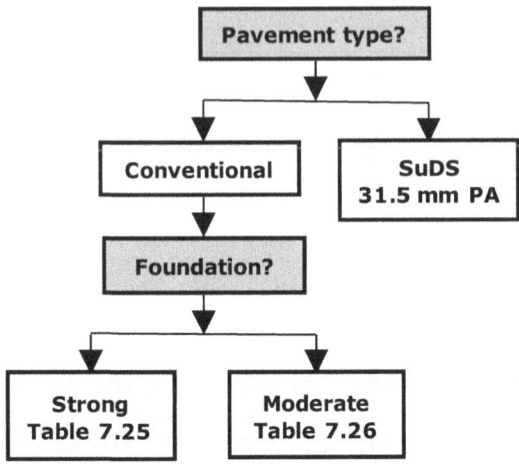

Figure 7.7 Flow-chart for base of roads.

Table 7.25 Mixtures for Base of Roads, Docks and Race Tracks over a Strong Foundation

Traffic	Cost band	Mixture type	Suggested NMAS	Design specifics
Heavy	Economy	AC/MA		–
		AC/HDM or HMB		–
	Standard	HRA	31.5 mm	–
	Premium	AC/SP		–
		AC/EME2		–
Medium or light	Economy	AC/MA		–
		AC/DBM		–
	Standard	HRA	31.5 mm	–
	Premium	AC/SP		–

Table 7.26 Mixtures for Base of Roads, Docks and Race Tracks over a Moderate Foundation

Traffic	Cost band	Mixture type	Suggested NMAS	Design specifics
Heavy	Economy	AC/MA	31.5 mm	–
		AC/HDM or HMB		–
	Standard	**HRA**		–
	Premium	AC/SP		–
Medium or light	Economy	AC/MA	31.5 mm	–
		AC/DBM		–
	Standard	HRA		–
	Premium	AC/SP		–

7.3 Airfields

7.3.1 Surface Course

A selection of suitable asphalt mixtures for use in the surface course of airfields is provided by following the flow-chart in Figure 7.8 to identify the relevant table of mixture type for the situation.

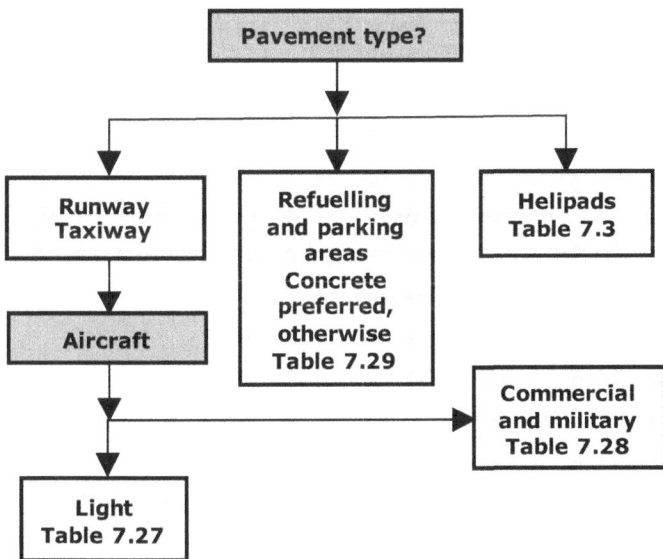

Figure 7.8 Flow-chart for surface course of airfields.

Table 7.27 Mixtures for Surface Course of Runway and Taxiway for Light Aircraft

Traffic	Cost band	Mixture type	Suggested NMAS	Design specifics
All	Economy	Grass	n/a	–
	Standard	**AC/DBM**		–
		AC/MA		–
		AC/SP	10 mm	–
	Premium	SMA		–
		BBTM		–

Table 7.28 Mixtures for Surface Course of Runway and Taxiway for Commercial and Military Aircraft

Traffic	Cost band	Mixture type	Suggested NMAS	Design specifics
Heavy	Economy	SMA	14 mm	–
		BBTM		–
	Standard	AC/MA	14 mm or 20 mm	–
		AC/BBA	14 mm	–
	Premium	AC/SP		–
		PA	10 mm	–
Medium or light	Economy	AC/DBM		–
		SMA	14 mm	–
		BBTM		–
	Standard	**AC/MA**	14 mm or 20 mm	–
		AC/BBA		–
		HRA	14 mm	–
	Premium	AC/SP		–
		PA	10 mm	–

Concrete preferred, but if asphalt is required:

Table 7.29 Mixtures for Surface Course of Refuelling and Parking Areas of Airfields and on Fuel Stations Alongside Roads

Traffic	Cost band	Mixture type	Suggested NMAS	Design specifics
Any	Standard	AC/MA	14 mm or 20 mm	Fuel-resistant binder
		HRA	14 mm	
	Premium	AC/SP		

Table 7.30 Mixtures for Surface Course of Helipads

Traffic	Cost band	Mixture type	Suggested NMAS	Design specifics
Heavy	Economy	SMA	14 mm	–
		BBTM		–
	Standard	AC/MA	14 mm or 20 mm	–
		AC/BBA		–
	Premium	AC/SP	14 mm	–
		AC/DBM		–
Medium or light	Economy	SMA		–
		BBTM	14 mm	–
	Standard	AC/MA	14 mm or 20 mm	–
		AC/BBA		–
	Premium	AC/SP	14 mm	–

7.3.2 Binder Course

A selection of suitable asphalt mixtures for use in the binder course of airfields is provided in Table 7.31.

Table 7.31 Mixtures for Binder Course on Airfields

Traffic	Cost band	Mixture type	Suggested NMAS	Design specifics
Heavy	Economy	AC/HDM or HMB	20 mm	
		AC/MA	20 mm or 31.5 mm	Design mixture
	Standard	**AC/BBA**		
		AC/SP		
		SMA	20 mm	
	Premium	AC/EME2		Strong foundation needed
		HRA		Recipe mixture
Medium or light	Economy	**AC/DBM**	20 mm	
		AC/MA	20 mm or 31.5 mm	
	Standard	AC/BBA		Recipe mixture
		SMA		
		AC/SP	20 mm	
	Premium	HRA		

7.3.3 Base

A selection of suitable asphalt mixtures for use in the base of airfields is provided in Table 7.32.

Table 7.32 Mixtures for Base on Airfields

Traffic	Cost band	Mixture type	Suggested NMAS	Design specifics
All	Economy	AC/HDM or HMB		
		AC/MA		Design mixture
	Standard	**AC/BBA**	31.5 mm	
		HRA		
		AC/SP		
	Premium	AC/EME2		Strong foundation needed

7.4 Footways and Cycleways

7.4.1 Surface Course

A selection of suitable asphalt mixtures for use in the surface course of foot-ways and cycleways is provided by following the flow-chart in Figure 7.4 if the pavement is regularly overrun by vehicles or in Figure 7.9 if not.

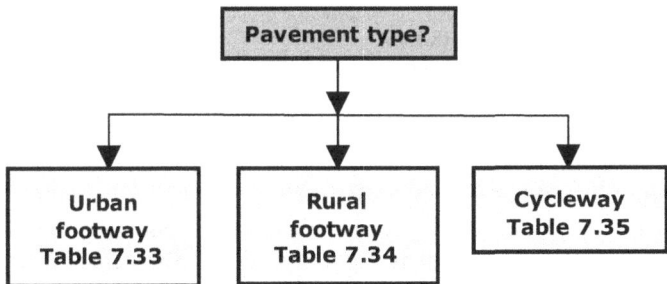

Figure 7.9 Flow-chart for surface course of footways and cycleways not regularly overrun by vehicles.

Table 7.33 Mixtures for Surface Course on Urban Footways not Regularly Overrun by Vehicles

Usage	Cost band	Mixture type	Suggested NMAS	Design specifics
	Economy	AC/DBM		–
		AC/MA		–
	Standard	SMA		–
All		BBTM	10 mm	–
		MA		–
	Premium	HRA		–

Table 7.34 Mixtures for Surface Course on Rural Footways not Regularly Overrun by Vehicles

Usage	Cost band	Mixture type	Suggested NMAS	Design specifics
	Economy	AC/OBM		–
	Standard	AC/MA		–
		AC/DBM		–
All		SMA	10 mm	–
	Premium	BBTM		–
		HRA		–

Table 7.35 Mixtures for Surface Course on Cycleways not Regularly Overrun by Vehicles

Usage	Cost band	Mixture type	Suggested NMAS	Design specifics
	Economy	AC/DBM		–
		AC/MA		–
	Standard	**SMA**		–
All		BBTM	10 mm	–
		MA		–
	Premium	HRA		–

7.4.2 Structural Layer(s)

A selection of suitable asphalt mixtures for use in the structural layer or layers of footways and cycleways is provided by following the flow-chart in Figure 7.6 if the pavement is regularly overrun by vehicles or in Table 7.36 if not.

Table 7.36 Mixtures for Structural Layer(s) on Footways and Cycleways not Regularly Overrun by Vehicles

Usage	Cost band	Mixture type	Suggested NMAS	Design specifics
	Economy	**AC/DBM**		–
	Standard	AC/MA		–
All		SMA	20 mm	–
	Premium	HRA		–

7.5 Parking Areas

7.5.1 Surface Course

A selection of suitable asphalt mixtures for use in the surface course of parking areas is provided by following the flow-chart in Figure 7.10 to identify the relevant table of mixture type for the situation.

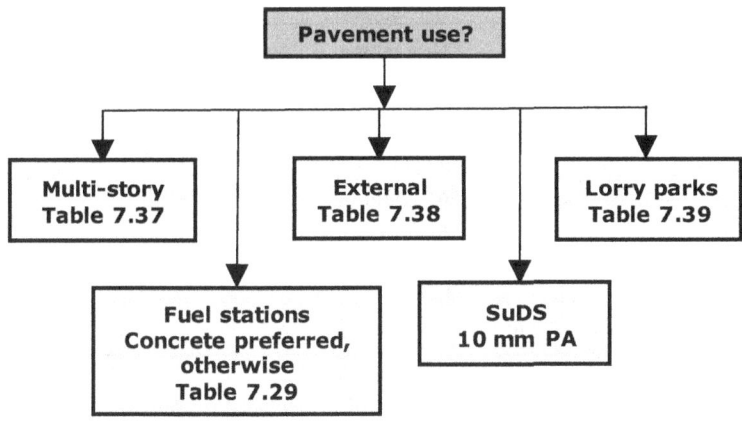

Figure 7.10 Flow-chart for surface course of parking areas.

Table 7.37 Mixtures for Surface Course of Multi-Storage Carparks

Usage	Cost band	Mixture type	Suggested NMAS	Design specifics
All	Economy	AC/DBM	10 mm	–
		AC/MA		–
	Standard	**SMA**		–
		BBTM		–
	Premium	HRA		–

Table 7.38 Mixtures for Surface Course of External Carparks and Playgrounds Regularly Overrun by Vehicles

Usage	Cost band	Mixture type	Suggested NMAS	Design specifics
Heavy	Economy	AC/DBM	10 mm	–
		AC/MA		–
	Standard	**SMA**		–
		BBTM		–
	Premium	HRA		–
Medium or light	Economy	AC/DBM	10 mm	–
		AC/OBM		–
		AC/MA		–
	Standard	**SMA**		–
		BBTM		–
	Premium	HRA		–

Table 7.39 Mixtures for Surface Course of Lorry Parks

Usage	Cost band	Mixture type	Suggested NMAS	Design specifics
		AC/MA		–
Heavy	Standard	**SMA**		–
		BBTM		–
	Economy	AC/DBM		–
		AC/MA		–
Medium or light	Standard	**SMA**	10 mm	–
		BBTM		–
	Premium	HRA		–

7.5.2 Structural Layers

The building structure is assumed to provide the structural layers of multi-storey carparks. A selection of suitable asphalt mixtures for the structural layers in other parking areas is provided by Figure 7.11.

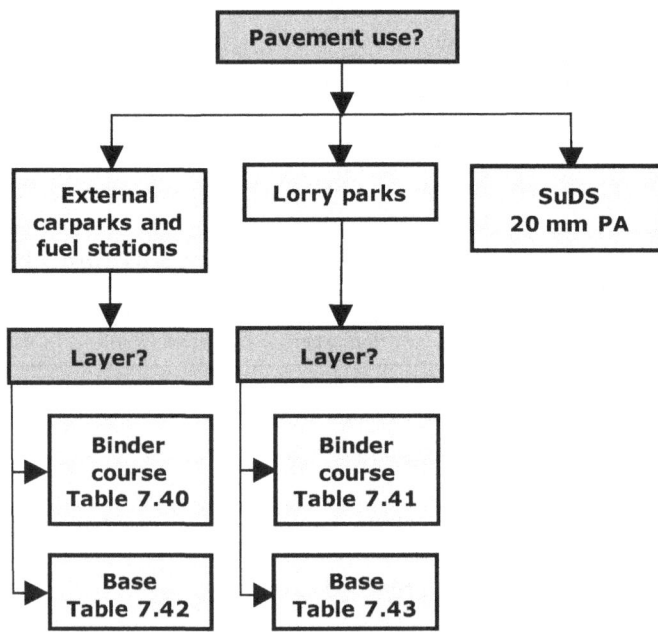

Figure 7.11 Flow-chart for structural layers of parking areas.

Table 7.40 Mixtures for Binder Course of External Carparks and Fuel Stations

Usage	Cost band	Mixture type	Suggested NMAS	Design specifics
All	Economy	**AC/DBM**		–
	Standard	AC/MA		–
		SMA	20 mm	–
	Premium	HRA		–

Table 7.41 Mixtures for Binder Course of Lorry Parks

Usage	Cost band	Mixture type	Suggested NMAS	Design specifics
All	Economy	**AC/HDM or HMB**		–
	Standard	AC/MA		–
		SMA	20 mm	–
		AC/SP		–
	Premium	HRA		–

Table 7.42 Mixtures for Base of External Carparks and Fuel Stations

Usage	Cost band	Mixture type	Suggested NMAS	Design specifics
All	Economy	AC/MA		–
		AC/DBM		–
	Standard	HRA	31.5 mm	–
	Premium	AC/SP		–

Table 7.43 Mixtures for Base of Lorry Parks

Usage	Cost band	Mixture type	Suggested NMAS	Design specifics
All	Economy	AC/MA		–
		AC/HDM or HMB		–
	Standard	**HRA**	31.5 mm	–
	Premium	AC/SP		–

7.6 Docks

A selection of suitable asphalt mixtures for use in the layers of docks is provided by following the flow-chart in Figure 7.12 to identify the relevant table of mixture type for the situation.

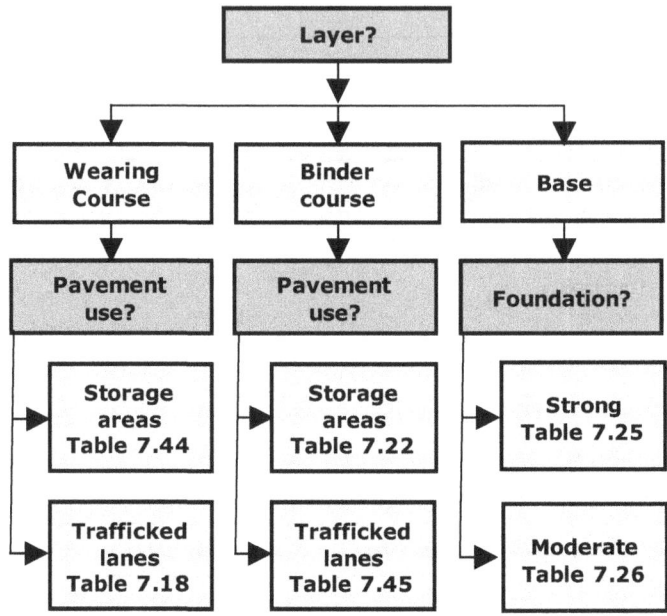

Figure 7.12 Flow-chart for all layers of docks.

Table 7.44 Mixtures for Surface Course of Dock Storage Areas

Usage	Cost band	Mixture type	Suggested NMAS	Design specifics
High	Standard	AC/MA		–
	Premium	AC/SP		–
		PA/GM		–
Medium or light	Standard	AC/MA		–
		BBTM	10 mm or 14 mm	–
		SMA		–
		AC/SP		–
	Premium	HRA+PCC		–
		PA/GM		–

Table 7.45 Mixtures for Binder Course of Dock Storage Areas

Usage	Cost band	Mixture type	Suggested NMAS	Design specifics
Heavy	Standard	AC/MA		Design mixture
		AC/SP		
	Premium	**AC/EME2**		Strong foundation needed
		HRA		Design mixture
Medium or light	Standard	AC/MA	20 mm	
		SMA		
		AC/SP		Recipe mixture
	Premium	**HRA**		

7.7 Ancillary Purposes

A selection of suitable asphalt mixtures for use on paved areas not covered in Sections 7.2 to 7.6 is provided by following the flow-chart in Figure 7.13 to identify the relevant table of mixture type for the situation.

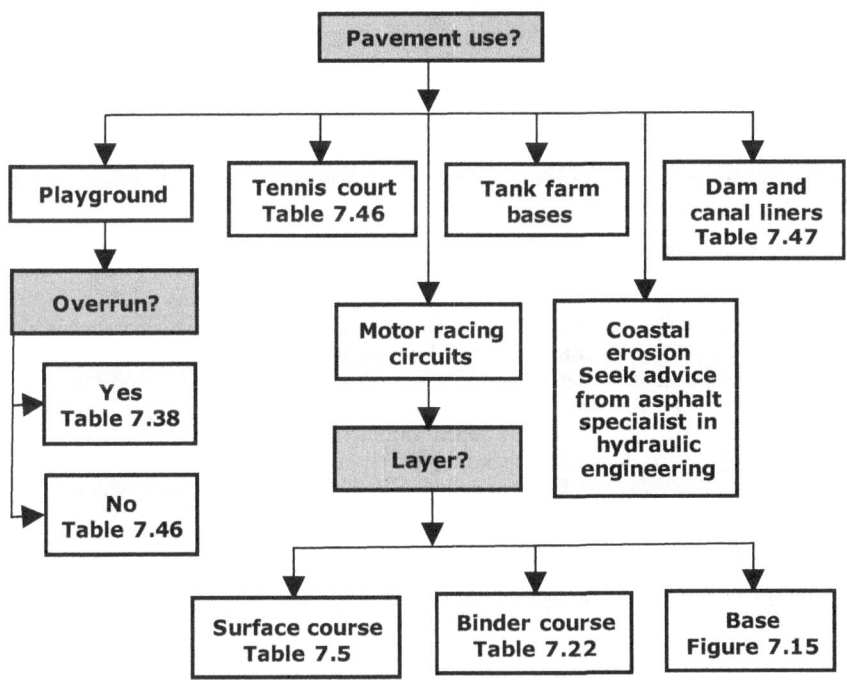

Figure 7.13 Flow-chart for surface course of ancillary purposes.

Table 7.46 Mixtures for Playgrounds not Regularly Overrun by Vehicles and Tennis Courts

Layer	Cost band	Mixture type	Suggested NMAS	Design specifics
Surface course	Economy	AC/OBM	10 mm	Rubber coarse aggregate preferred
	Standard	**AC/DBM**		
		SMA		
		BBTM		
	Premium	AC/MA		
		MA		
Structural layer(s)	Economy	**AC/DBM**	20 mm	–
	Standard	AC/MA		–
		AC/HDM or HMB		–
		SMA		–

Table 7.47 Mixtures for Tanking Farm Bases and for Dam and Canal Liners

Usage	Cost band	Mixture type	Suggested NMAS	Design specifics
All	Economy	HRA	10 mm	–
	Standard	**MA**		–
		MA/Guss		–

Note

1   In the tables and figures, NMAS is the nominal maximum aggregate sizes and the mixtures are: AC/BBA = béton bitumineux pour chaussées aéronautiques; AC/DBM = dense bituminous macadam; AC/EME = enrobés à module élevé; AC/HDM = heavy duty macadam; AC/HMB = high modulus base; AC/MA = Marshall asphalt; AC/OBM = open bitumen macadam; AC/SP = superpave asphalt; AUTL = asphalt for ultra-thin layers; HRA/HSC = high stone content hot rolled asphalt; HRA+PCC = hot rolled asphalt with pre-coated chippings; HRA/SC = sand carpet; MA = mastic asphalt; MA/Gus = gussasphalt; MA+PCC = mastic asphalt with pre-coated chippings; PA = porous asphalt; PA/GM = grouted macadam; PA/TL = twin layer porous asphalt; SA = soft asphalt; SMA = stone mastic asphalt.

Chapter 8

Uses for Asphalt Mixtures

8.1 General Approach

This chapter lists the uses that different mixtures are recommended for in Table 7.1 to Table 7.47 together with those referred to directly in Figure 7.6, Figure 7.7, Figure 7.10, Figure 7.11 and Figure 7.13. As such, it is a useful reorganisation of Chapter 7 from the reverse direction (sites where each asphalt mixture can be used successfully rather than the asphalt mixtures that can be used at each site) rather than anything new in itself.

8.2 Asphalt Concrete

6 mm (or similar size) Marshall asphalt (AC/MA) is suitable for use as the material for:

- Surface course of residential roads (Table 7.17), and
- Surface course of long-span bridge decks (Table 7.20).

10 mm (or similar size) AC/MA is suitable for use as the material for:

- Surface course of dual-carriageway roads on bends or slopes (Table 7.3), including in damp conditions (Table 7.4), and at roundabouts, including in damp conditions (Table 7.5),
- Surface course of single-carriageway urban roads on bends or slopes (Table 7.8), including in damp conditions (Table 7.9), and at round-abouts including in damp conditions (Table 7.5),
- Surface course of rural single-carriageway roads on roundabouts, sharp bends or steep slopes including in damp conditions (Table 7.15) and snow-bound conditions (Table 7.16),
- Surface course of residential roads (Table 7.17),
- Surface course of industrial roads (Table 7.18),
- Surface course of long tunnels (Table 7.19),
- Surface course of long-span bridge decks (Table 7.20),

- Layer above a waterproofing layer of roads over bridges and similar structures (Table 7.24),
- Surface course of runway and taxiway for light aircraft (Table 7.27),
- Surface course on urban (Table 7.33) and rural (Table 7.34) footways not regularly overrun by vehicles,
- Surface course on cycleways not regularly overrun by vehicles (Table 7.35),
- Surface course of multi-storey carparks (Table 7.37), external carparks (Table 7.38) and lorry parks (Table 7.39),
- Surface course of trafficked lanes (Table 7.18) and storage areas (Table 7.44) of ports,
- Surface course for motor racing circuits (Table 7.5), and
- Surface course for playgrounds regularly overrun by vehicles (Table 7.38) and not regularly overrun by vehicles (Table 7.46),
- Surface course for tennis courts (Table 7.46).

14 mm (or similar size) AC/MA is suitable for use as the material for:

- Surface course of straight and level dual-carriageway roads (Table 7.1) including in damp conditions (Table 7.2),
- Surface course of straight and level urban single-carriageway roads (Table 7.6) including in damp conditions (Table 7.7),
- Surface course of straight and level rural single-carriageway roads (Table 7.10) including in damp conditions (Table 7.11) and snow-bound conditions (Table 7.12),
- Surface course of rural single-carriageway roads on bends or slopes including in damp conditions (Table 7.13) or snow-bound conditions (Table 7.14),
- Surface course of industrial roads (Table 7.18),
- Surface course of long tunnels (Table 7.19),
- Surface course of runway and taxiway for commercial and military aircraft (Table 7.28),
- Surface course of refuelling and parking areas of airfields if concrete cannot be used (Table 7.29),
- Surface course of helipads (Table 7.30),
- Surface course of fuel stations alongside roads if concrete cannot be used (Table 7.29), and
- Surface course of trafficked lanes (Table 7.18) and storage areas (Table 7.44) of ports.

20 mm (or similar size) AC/MA is suitable for use as the material for:

- Binder course of low (Table 7.21) or high (Table 7.22) stressed roads with conventional surface course,

- Binder course of conventional roads with permeable surface course (Table 7.23),
- Surface course of runway and taxiway for commercial and military aircraft (Table 7.28),
- Surface course of refuelling and parking areas of airfields if concrete cannot be used (Table 7.29),
- Surface course of helipads (Table 7.30),
- Binder course on airfields (Table 7.31),
- Surface course of fuel stations alongside roads if concrete cannot be used (Table 7.29),
- Structural layer(s) on footways and cycleways not regularly overrun by vehicles (Table 7.36),
- Binder course of external carparks and fuel stations (Table 7.40),
- Binder course of lorry parks (Table 7.41),
- Binder course of trafficked lanes (Table 7.22) and storage areas (Table 7.45) of docks,
- Binder course of motor racing circuits (Table 7.22), and
- Structural layer(s) for playgrounds not regularly overrun by vehicles (Table 7.46).
- Structural layer(s) for tennis courts (Table 7.46).

31.5 mm (or similar size) AC/MA is suitable for use as the material for:

- Base of roads over a strong (Table 7.25) or moderate foundation (Table 7.26),
- Base on airfields (Table 7.32),
- Base of lorry parks (Table 7.43),
- Base of docks over a strong (Table 7.25) or moderate foundation (Table 7.26), and
- Base of motor racing circuits over a strong (Table 7.25) or moderate foundation (Table 7.26).

10 mm (or similar size) Superpave asphalt (AC/SP) is suitable for use as the material for:

- Surface course of dual-carriageway roads on bends or slopes (Table 7.3), including in damp conditions (Table 7.4), and at roundabouts, including in damp conditions (Table 7.5),
- Surface course of single-carriageway urban roads on bends or slopes (Table 7.8), including in damp conditions (Table 7.9), and at round-abouts, including in damp conditions (Table 7.5),
- Surface course of industrial roads (Table 7.18),
- Surface course of long tunnels (Table 7.19),
- Surface course of long-span bridge decks (Table 7.20),

- Surface course of runway and taxiway for light aircraft (Table 7.27),
- Surface course of trafficked lanes (Table 7.18) and storage areas (Table 7.44) of ports, and
- Surface course of motor racing circuits (Table 7.5).

14 mm (or similar size) AC/SP is suitable for use as the material for:

- Surface course of straight and level dual-carriageway roads (Table 7.1) including in damp conditions (Table 7.2),
- Surface course of straight and level urban single-carriageway roads (Table 7.6) including in damp conditions (Table 7.7),
- Surface course of industrial roads (Table 7.18),
- Surface course of long tunnels (Table 7.19),
- Surface course of runway and taxiway for commercial and military aircraft (Table 7.28),
- Surface course of refuelling and parking areas of airfields if concrete cannot be used (Table 7.29),
- Surface course of helipads (Table 7.30),
- Surface course of fuel stations alongside roads if concrete cannot be used (Table 7.29), and
- Surface course of trafficked lanes (Table 7.18) and storage areas (Table 7.44) of ports.

20 mm (or similar size) AC/SP is suitable for use as the material for:

- Binder course of low stressed roads with heavy traffic (Table 7.21) or high stressed roads with any traffic (Table 7.22), each with conventional surface course,
- Binder course of conventional roads with permeable surface course (Table 7.23),
- Binder course on airfields (Table 7.31),
- Binder course of lorry parks (Table 7.41),
- Binder course of trafficked lanes (Table 7.22) and storage areas (Table 7.45) of docks, and
- Binder course of motor racing circuits (Table 7.22).

31.5 mm (or similar size) AC/SP is suitable for use as the material for:

- Base of roads over a strong (Table 7.25) or moderate foundation (Table 7.26),
- Base on airfields (Table 7.32),
- Base of lorry parks (Table 7.43),
- Base of docks over a strong (Table 7.25) or moderate foundation (Table 7.26), and

- Base of motor racing circuits over a strong (Table 7.25) or moderate foundation (Table 7.26).

6 mm (or similar size) dense bitumen macadam (AC/DBM) is suitable for use as the material for:

- Surface course of residential roads (Table 7.17),
- Surface course of long-span bridge decks (Table 7.20), and
- Layer above a waterproofing layer of roads over bridges and similar structures (Table 7.24).

10 mm (or similar size) AC/DBM is suitable for use as the material for:

- Surface course of urban single carriageway roads on bends or slopes (Table 7.8), including in damp conditions (Table 7.9),
- Surface course of residential roads (Table 7.17),
- Surface course of industrial roads with medium or light traffic (Table 7.18),
- Surface course of long-span bridge decks (Table 7.20),
- Layer above a waterproofing layer of roads over bridges and similar structures (Table 7.24),
- Surface course of runway and taxiway for light aircraft (Table 7.27),
- Surface course of runway and taxiway for commercial and military aircraft with medium or light traffic (Table 7.28),
- Surface course on urban footways not regularly overrun by vehicles (Table 7.33),
- Surface course on cycleways not regularly overrun by vehicles (Table 7.35),
- Surface course of lorry parks with medium or light traffic (Table 7.39), multi-storey carparks (Table 7.37) and external carparks (Table 7.38),
- Surface course of trafficked lanes of ports (Table 7.18),
- Surface course for playgrounds not regularly overrun by vehicles (Table 7.46), and
- Surface course for tennis courts (Table 7.46).

14 mm (or similar size) AC/DBM is suitable for use as the material for:

- Surface course of straight and level urban single-carriageway roads (Table 7.6) including in damp conditions (Table 7.7),
- Surface course of straight and level rural single-carriageway roads (Table 7.10) including in damp conditions (Table 7.11) and snow-bound conditions (Table 7.12),

- Surface course of rural single-carriageway roads on bends or slopes including in damp conditions (Table 7.13) or snow-bound conditions (Table 7.14),
- Surface course of industrial roads with medium or light traffic (Table 7.18),
- Surface course of helipads with medium or light traffic (Table 7.30), and
- Surface course of trafficked lanes of ports (Table 7.18).

20 mm (or similar size) AC/DBM is suitable for use as the material for:

- Binder course of low (Table 7.21) or high (Table 7.22) stressed roads with conventional surface course with medium or light traffic,
- Binder course on airfields with medium or light traffic (Table 7.31),
- Structural layer(s) on footways and cycleways not regularly overrun by vehicles (Table 7.36),
- Binder course of external carparks and fuel stations (Table 7.40),
- Binder course of trafficked lanes of docks with medium or light traffic (Table 7.22),
- Binder course of motor racing circuits (Table 7.22),
- Structural layer(s) for playgrounds not regularly overrun by vehicles (Table 7.46), and
- Structural layer(s) for tennis courts (Table 7.46).

31.5 mm (or similar size) AC/DBM is suitable for use as the material for:

- Base of roads over a strong (Table 7.25) or moderate foundation (Table 7.26) with medium or light traffic,
- Base of docks over a strong (Table 7.25) or moderate foundation (Table 7.26), and
- Base of motor racing circuits over a strong (Table 7.25) or moderate foundation (Table 7.26).

20 mm (or similar size) heavy duty mixture (AC/HDM) and high modulus base (AC/HMB) are suitable for use as:

- Binder course of low (Table 7.21) or high (Table 7.22) stressed roads with conventional surface course with heavy traffic,
- Binder course on airfields with heavy traffic (Table 7.31),
- Binder course of lorry parks (Table 7.41),
- Binder course of trafficked lanes of docks with heavy traffic (Table 7.22),
- Binder course of motor racing circuits (Table 7.22), and
- Structural layer(s) for playgrounds not regularly overrun by vehicles (Table 7.46), and

- Structural layer(s) for tennis courts (Table 7.46).

31.5 mm (or similar size) AC/HDM and AC/HMB are suitable for use as:

- Base of roads over a strong (Table 7.25) or moderate foundation (Table 7.26) with heavy traffic,
- Base of docks over a strong (Table 7.25) or moderate foundation (Table 7.26) with heavy traffic,
- Base on airfields (Table 7.32),
- Base of lorry parks (Table 7.43),
- Base of docks over a strong (Table 7.25) or moderate foundation (Table 7.26) with heavy traffic, and
- Base of motor racing circuits over a strong (Table 7.25) or moderate foundation (Table 7.26) with heavy traffic.

10 mm (or similar size) open bitumen macadams (AC/OBM) is suitable for use as the material for:

- Surface course on rural footways not regularly overrun by vehicles (Table 7.34),
- Surface course of external carparks with medium or light traffic (Table 7.38), and
- Surface course for playgrounds regularly overrun by vehicles (Table 7.38) and not regularly overrun by vehicles (Table 7.46),
- Surface course for tennis courts (Table 7.46).

14 mm (or similar size) AC/OBM is suitable for use as the material for:

- Surface course of straight and level rural single-carriageway roads with light but not slow-moving traffic (Table 7.10) but not in damp conditions (Table 7.11) or snow-bound conditions (Table 7.12).

20 mm (or similar size) enrobés à module élevé (AC/EME) is suitable for use as the material for:

- Binder course of low (Table 7.21) or high (Table 7.22) stressed roads with conventional surface course with heavy traffic,
- Binder course of conventional roads with permeable surface course with heavy traffic (Table 7.23),
- Binder course on airfields with heavy traffic and strong foundations (Table 7.31),
- Binder course of trafficked lanes (Table 7.22) and storage areas (Table 7.45) of docks with heavy usage and strong foundations, and

- Binder course of motor racing circuits (Table 7.22) with strong foundations.

31.5 mm (or similar size) AC/EME is suitable for use as the material for:

- Base of roads with heavy traffic over a strong (Table 7.25) but not moderate foundation (Table 7.26),
- Base on airfields over a strong foundation (Table 7.32),
- Base of docks with heavy traffic over a strong (Table 7.25) but not moderate foundation (Table 7.26), and
- Base of motor racing circuits over a strong (Table 7.25) or moderate foundation (Table 7.26) with heavy traffic.

14 mm (or similar size) béton bitumineux pour chaussées aéronautiques (AC/BBA) is suitable for use as the material for:

- Surface course of runway and taxiway for commercial and military aircraft (Table 7.28); and
- Surface course of helipads (Table 7.30).

20 mm (or similar size) AC/BBA is suitable for use as the material for:

- Binder course on airfields (Table 7.31).

31.5 mm (or similar size) AC/BBA is suitable for use as the material for:

- Base on airfields (Table 7.32).

8.3 Stone Mastic Asphalt

6 mm (or similar size) stone mastic asphalt (SMA) is suitable for use as the material for:

- Surface course of residential roads (Table 7.17), and
- Surface course of long-span bridge decks (Table 7.20).

10 mm (or similar size) SMA is suitable for use as the material for:

- Surface course of dual-carriageway roads on bends or slopes (Table 7.3), including in damp conditions (Table 7.4), and at roundabouts, including in damp conditions (Table 7.5),
- Surface course of single-carriageway urban roads on bends or slopes (Table 7.8), including in damp conditions (Table 7.9), and at roundabouts, including in damp conditions (Table 7.5),

- Surface course of rural single-carriageway roads on roundabouts, sharp bends or steep slopes including in damp conditions (Table 7.15) and snow-bound conditions (Table 7.16),
- Surface course of residential roads (Table 7.17),
- Surface course of industrial roads (Table 7.18),
- Surface course of long tunnels (Table 7.19),
- Surface course of long-span bridge decks (Table 7.20),
- Layer above a waterproofing layer of roads over bridges and similar structures (Table 7.24),
- Surface course of runway and taxiway for light aircraft (Table 7.27),
- Surface course on urban (Table 7.33) and rural (Table 7.34) footways not regularly overrun by vehicles,
- Surface course on cycleways not regularly overrun by vehicles (Table 7.35),
- Surface course of multi-storey carparks (Table 7.37), external carparks (Table 7.38) and lorry parks (Table 7.39),
- Surface course of trafficked lanes (Table 7.18) and storage areas with medium or light loading (Table 7.44) of ports,
- Surface course of motor racing circuits (Table 7.5), and
- Surface course for playgrounds regularly overrun by vehicles (Table 7.38) and not regularly overrun by vehicles (Table 7.46),
- Surface course for tennis courts (Table 7.46).

14 mm (or similar size) SMA is suitable for use as the material for:

- Surface course of straight and level dual-carriageway roads (Table 7.1) including in damp conditions (Table 7.2),
- Surface course of straight and level urban single-carriageway roads (Table 7.6) including in damp conditions (Table 7.7),
- Surface course of straight and level rural single-carriageway roads (Table 7.10) including in damp conditions (Table 7.11) and snow-bound conditions (Table 7.12),
- Surface course of rural single-carriageway roads on bends or slopes including in damp conditions (Table 7.13) or snow-bound conditions (Table 7.14),
- Surface course of industrial roads (Table 7.18),
- Surface course of long tunnels (Table 7.19),
- Surface course of runway and taxiway for commercial and military aircraft (Table 7.28),
- Surface course of helipads (Table 7.30), and
- Surface course of trafficked lanes (Table 7.18) and storage areas with medium or light loading (Table 7.44) of ports.

20 mm (or similar size) SMA is suitable for use as the material for:

- Binder course of low (Table 7.21) or high (Table 7.22) stressed roads with conventional surface course,
- Binder course of conventional roads with permeable surface course with light or medium traffic (Table 7.23),
- Binder course on airfields (Table 7.31),
- Structural layer(s) on footways and cycleways not regularly overrun by vehicles (Table 7.36),
- Binder course of external carparks and fuel stations (Table 7.40),
- Binder course of lorry parks (Table 7.41),
- Binder course of trafficked lanes (Table 7.22) and storage areas with medium or light usage (Table 7.45) of docks,
- Binder course of motor racing circuits (Table 7.22),
- Structural layer(s) for playgrounds not regularly overrun by vehicles (Table 7.46), and
- Structural layer(s) for tennis courts (Table 7.46).

8.4 Asphalt Concrete for Very Thin Layers

6 mm (or similar size) asphalt concrete for very thin layers (BBTM) is suitable for use as the material for:

- Surface course of residential roads (Table 7.17),
- Surface course of long-span bridge decks (Table 7.20), and
- Layer above a waterproofing layer of roads over bridges and similar structures (Table 7.24).

10 mm (or similar size) BBTM is suitable for use as the material for:

- Surface course of dual-carriageway roads on bends or slopes (Table 7.3), including in damp conditions (Table 7.4), and at roundabouts, including in damp conditions (Table 7.5),
- Surface course of single-carriageway urban roads on bends or slopes (Table 7.8), including in damp conditions (Table 7.9), and at round-abouts, including in damp conditions (Table 7.5),
- Surface course of rural single-carriageway roads on roundabouts, sharp bends or steep slopes including in damp conditions (Table 7.15) and snow-bound conditions (Table 7.16),
- Surface course of residential roads (Table 7.17),
- Surface course of industrial roads (Table 7.18),
- Surface course of long tunnels (Table 7.19),
- Surface course of long-span bridge decks (Table 7.20),
- Layer above a waterproofing layer of roads over bridges and similar structures (Table 7.24),
- Surface course of runway and taxiway for light aircraft (Table 7.27),

- Surface course on urban (Table 7.33) and rural (Table 7.34) footways not regularly overrun by vehicles,
- Surface course on cycleways not regularly overrun by vehicles (Table 7.35),
- Surface course of multi-storey carparks (Table 7.37), external car-parks (Table 7.38) and lorry parks (Table 7.39),
- Surface course of trafficked lanes (Table 7.18) and storage areas with medium or light loading (Table 7.44) of ports,
- Surface course of motor racing circuits (Table 7.5)
- Surface course for playgrounds regularly overrun by vehicles (Table 7.38) and not regularly overrun by vehicles (Table 7.46), and
- Surface course for tennis courts (Table 7.46).

14 mm (or similar size) BBTM is suitable for use as the material for:

- Surface course of straight and level dual-carriageway roads (Table 7.1) including in damp conditions (Table 7.2),
- Surface course of straight and level urban single-carriageway roads (Table 7.6) including in damp conditions (Table 7.7),
- Surface course of straight and level rural single-carriageway roads (Table 7.10) including in damp conditions (Table 7.11) and snow-bound conditions (Table 7.12),
- Surface course of rural single-carriageway roads on bends or slopes including in damp conditions (Table 7.13) or snow-bound conditions (Table 7.14),
- Surface course of industrial roads (Table 7.18),
- Surface course of long tunnels (Table 7.19),
- Surface course of runway and taxiway for commercial and military aircraft (Table 7.28),
- Surface course of helipads (Table 7.30), and
- Surface course of trafficked lanes (Table 7.18) and storage areas with medium or light loading (Table 7.44) of ports.

8.5 Asphalt for Ultra-Thin Layers

10 mm (or similar size) asphalt for ultra-thin layers (AUTL) is suitable for use as the material for:

- Surface course of dual-carriageway roads on bends or slopes but not with heavy traffic (Table 7.3) but not in damp conditions (Table 7.4),
- Surface course of single-carriageway urban roads on bends or slopes (Table 7.8) including in damp conditions (Table 7.9), and
- Surface course of long-span bridge decks (Table 7.20).14 mm (or similar size) AUTL is suitable for use as the material for:

- Surface course of straight and level dual-carriageway roads (Table 7.1) but not in damp conditions (Table 7.2),
- Surface course of straight and level urban single-carriageway roads (Table 7.6) but not in damp conditions (Table 7.7),
- Surface course of straight and level rural single-carriageway roads with light or medium but not slow-moving traffic (Table 7.10) including in damp conditions (Table 7.11) and snow-bound conditions (Table 7.12), and
- Surface course of rural single-carriageway roads on bends or slopes including in damp conditions (Table 7.13) or snow-bound conditions (Table 7.14).

8.6 Porous Asphalt

10 mm (or similar size) single layer porous asphalt (PA) is suitable for use as the material for:

- Surface course of straight and level dual-carriageway roads (Table 7.1), including in damp conditions (Table 7.2), and on bends or slopes (Table 7.3), including in damp conditions (Table 7.4),
- Surface course of straight and level urban single-carriageway roads (Table 7.6), including in damp conditions (Table 7.7), and on bends or slopes (Table 7.8), including in damp conditions (Table 7.9),
- Surface course of runway and taxiway for commercial and military aircraft (Table 7.28), and
- Surface course of parking areas as part of sustainable drainage systems (SuDS: Figure 7.10).

20 mm (or similar size) single layer PA is suitable for use as the material for:

- Surface course of straight and level dual-carriageway roads (Table 7.1), including in damp conditions (Table 7.2), and on bends or slopes (Table 7.3), including in damp conditions (Table 7.4),
- Surface course of straight and level urban single-carriageway roads (Table 7.6), including in damp conditions (Table 7.7), and on bends or slopes (Table 7.8), including in damp conditions (Table 7.9),
- Binder course for roads as part of SuDS (Figure 7.6),
- Binder course of low stressed roads with conventional surface course with medium or light traffic but only when groundwater cannot be excluded from pavement structure (Table 7.21), and
- Binder course of parking areas as part of SuDS (Figure 7.11).

31.5 mm (or similar size) single layer PA is suitable for use as the material for:

- Base for roads as part of SuDS (Figure 7.7), and
- Base of parking areas as part of SuDS (Figure 7.11).

10 mm and 20 mm (or similar sizes) twin layer porous asphalt (PA/TL) is suitable for use as the material for:

- Surface course of straight and level dual-carriageway roads (Table 7.1), including in damp conditions (Table 7.2), and on bends or slopes (Table 7.3), including in damp conditions (Table 7.4), and
- Surface course of straight and level urban single-carriageway roads (Table 7.6), including in damp conditions (Table 7.7), and on bends or slopes (Table 7.8), including in damp conditions (Table 7.9).

10 mm (or similar size) grouted macadam (PA/GM) is suitable for use as the material for:

- Surface course of industrial roads (Table 7.18), and
- Surface course of trafficked lanes (Table 7.18) and storage areas (Table 7.44) of ports.

14 mm (or similar size) PA/GM is suitable for use as the material for:

- Surface course of industrial roads (Table 7.18), and
- Surface course of trafficked lanes (Table 7.18) and storage areas loading (Table 7.44) of ports.

8.7 Mastic Asphalt

10 mm (or similar size) mastic asphalt (MA) is suitable for use as the material for:

- Layer above a waterproofing layer of roads over bridges and similar structures (Table 7.24),
- Surface course on urban footways not regularly overrun by vehicles (Table 7.33),
- Surface course on cycleways not regularly overrun by vehicles (Table 7.35),
- Surface course for playgrounds not regularly overrun by vehicles (Table 7.46), and
- Surface course for tennis courts (Table 7.46),
- Tanking farm bases (Table 7.47), and
- Dam and canal liners (Table 7.47).

10 mm (or similar size) mastic asphalt plus pre-coated chippings (MA +PCC) is suitable for use as the material for:

- Surface course of long tunnels (Table 7.19), and
- Surface course of long-span bridge decks (Table 7.20).

10 mm (or similar size) gussasphalt (MA/Guss) is suitable for use as the material for:

- Surface course of long tunnels (Table 7.19),
- Surface course of long-span bridge decks (Table 7.20),
- Tanking farm bases (Table 7.47), and
- Dam and canal liners (Table 7.47).

8.8 Hot Rolled Asphalt

6 mm (or similar size) high stone content hot rolled asphalt (HRA/HSC) is suitable for use as the material for:

- Surface course of residential roads (Table 7.17), and
- Layer above a waterproofing layer of roads over bridges and similar structures (Table 7.24).

10 mm (or similar size) HRA/HSC is suitable for use as the material for:

- Surface course of rural single-carriageway roads on roundabouts, sharp bends or steep slopes including in damp conditions (Table 7.15) and snow-bound conditions (Table 7.16),
- Surface course of residential roads (Table 7.17),
- Layer above a waterproofing layer of roads over bridges and similar structures (Table 7.24), and
- Surface course on urban (Table 7.33) and rural (Table 7.34) footways not regularly overrun by vehicles.
- Surface course on cycleways not regularly overrun by vehicles (Table 7.35),
- Surface course of lorry parks with medium or light traffic (Table 7.39), multi-storey carparks (Table 7.37) and external carparks (Table 7.38),
- Surface course for playgrounds regularly overrun by vehicles (Table 7.38) and not regularly overrun by vehicles (Table 7.46),
- Surface course for tennis courts (Table 7.46).
- Tanking farm bases (Table 7.47), and
- Dam and canal liners (Table 7.47).

14 mm (or similar size) HRA/HSC is suitable for use as the material for:

- Surface course of straight and level rural single-carriageway roads (Table 7.10) including in damp conditions (Table 7.11) and snow-bound conditions (Table 7.12),
- Surface course of rural single-carriageway roads on bends or slopes including in damp conditions (Table 7.13) or snow-bound conditions (Table 7.14),
- Surface course of runway and taxiway for commercial and military aircraft (Table 7.28),
- Surface course of refuelling and parking areas of airfields if concrete cannot be used (Table 7.29), and
- Surface course of fuel stations alongside roads if concrete cannot be used (Table 7.29).

20 mm (or similar size) HRA/HSC is suitable for use as the material for:

- Binder course of low (Table 7.21) or high (Table 7.22) stressed roads with conventional surface course,
- Binder course of conventional roads with permeable surface course (Table 7.23),
- Binder course on airfields (Table 7.31),
- Structural layer(s) on footways and cycleways not regularly overrun by vehicles (Table 7.36),
- Binder course of external carparks and fuel stations (Table 7.40),
- Binder course of lorry parks (Table 7.41),
- Binder course of trafficked lanes (Table 7.22) and storage areas (Table 7.45) of docks, and
- Binder course of motor racing circuits (Table 7.22).

31.5 mm (or similar size) HRA/HSC is suitable for use as the material for:

- Base of roads over a strong (Table 7.25) or moderate foundation (Table 7.26),
- Base on airfields (Table 7.32),
- Base of lorry parks (Table 7.43),
- Base of docks over a strong (Table 7.25) or moderate foundation (Table 7.26), and
- Base of motor racing circuits over a strong (Table 7.25) or moderate foundation (Table 7.26) with heavy traffic.

10 mm (or similar size) hot rolled asphalt plus pre-coated chippings (HRA+PCC) is suitable for use as the material for:

- urface course of dual-carriageway roads on bends or slopes (Table 7.3), including in damp conditions (Table 7.4), and at round-abouts, including in damp conditions (Table 7.5),
- Surface course of single-carriageway urban roads on bends or slopes (Table 7.8), including in damp conditions (Table 7.9), and at round-abouts, including in damp conditions (Table 7.5),
- Surface course of rural single-carriageway roads on roundabouts, sharp bends or steep slopes including in damp conditions (Table 7.15) and snow-bound conditions (Table 7.16),
- Surface course of industrial roads with medium or light traffic (Table 7.18),
- Surface course of long tunnels (Table 7.19),
- Surface course racing circuits (Table 7.5),
- Surface course of long-span bridge decks (Table 7.20), and
- Surface course of trafficked lanes (Table 7.18) and storage areas with medium or light loading (Table 7.44) of ports.

14 mm (or similar size) HRA+PCC) is suitable for use as the material for:

- Surface course of straight and level dual-carriageway roads (Table 7.1) including in damp conditions (Table 7.2),
- Surface course of straight and level urban single-carriageway roads (Table 7.6) including in damp conditions (Table 7.7),
- Surface course of straight and level rural single-carriageway roads (Table 7.10) including in damp conditions (Table 7.11) and snow-bound conditions (Table 7.12),
- Surface course of rural single-carriageway roads on bends or slopes including in damp conditions (Table 7.13) or snow-bound conditions (Table 7.14),
- Surface course of industrial roads with medium or light traffic (Table 7.18),
- Surface course of long tunnels (Table 7.19), and
- Surface course of trafficked lanes (Table 7.18) and storage areas with medium or light loading (Table 7.44) of ports.

Sand carpet (HRA/SC) has been applied, but is not recommended, for use in:

- Layer above a waterproofing layer of roads over bridges and similar structures (Table 7.24).

8.9 Soft Asphalt

10 mm (or similar size) soft asphalt (SA) is suitable for use as the material for:

- Surface course of rural single-carriageway roads with light and not low-speed traffic on roundabouts, sharp bends or steep slopes including snow-bound conditions (Table 7.16).

14 mm (or similar size) soft asphalt (SA) is suitable for use as the material for:

- Surface course of rural single-carriageway roads with light and not low-speed traffic in snow-bound conditions whether straight and level (Table 7.12) or on bends or slopes (Table 7.14).

Chapter 9

Examples and Summary

9.1 Examples

9.1.1 Major Road Example

The road type is dual carriageway, so go to Figure 7.3 for the surface course, Figure 7.6 for the binder course and Figure 7.7 for the base.

The geometry is straight and level with a normal climate, so the surface course is selected from Table 7.1. The preferred option would be 14 mm stone mastic asphalt (SMA), but the road authority has limited funds for the project, so 14 mm asphalt for ultra-thin layers (AUTL) is selected for the surface course.

The pavement is being designed as impermeable without a waterproofing layer and standard stress level, so the binder course is selected from Table 7.21. The traffic is heavy, so the preferred option would be 20 mm enrobé à module élevé type 2 (EME2), but the road authority has limited funds for the project, so 20 mm heavy duty mixture (HDM) is selected for the binder course.

The pavement is being designed as conventional over a strong foundation, so the base is selected from Table 7.25. The traffic is heavy, so the preferred option would be 31.5 mm EME2, but the road authority has limited funds for the project, so 31.5 mm HDM is selected for the base.

The choices of materials for this pavement are:

- Surface course – 14 mm AUTL.
- Binder course – 20 mm HDM.
- Base – 31.5 mm HDM.

9.1.2 Minor Road Example

The road type is rural single carriageway, so go to Figure 7.5 for the surface course, Figure 7.6 for the binder course and Figure 7.7 for the base.

The geometry includes a series of roundabouts in an area that is regularly snow-bound, so the surface course is selected from Table 7.16. The traffic level is light and the speed is slow, so the preferred option would be 10 mm asphalt concrete for very thin layers (BBTM), which is selected for the surface course.

The pavement is being designed as impermeable without a waterproofing layer and high stress level, so the binder course is selected from Table 7.22. The traffic is light, so the preferred option would be 20 mm dense bituminous macadam (AC/DBM), which is selected for the binder course.

The pavement is being designed as conventional over a weak foundation, so the base is selected from Table 7.26. The traffic is light, so the preferred option would be 31.5 mm DBM, but the route is strategically important so 31.5 mm hot rolled asphalt (HRA) is selected for the base.

The choices of materials for this pavement are:

- Surface course – 10 mm BBTM.
- Binder course – 20 mm AC/DBM.
- Base – 31.5 mm HRA.

9.1.3 Airfield Example

For airfield pavements, go to Figure 7.8 for the surface course, Table 7.31 for the binder course and Table 7.32 for the base.

The airfield type is a taxiway for commercial aircraft, so the surface course is selected from Table 7.28. The traffic level is medium, so the preferred option would be 14 mm or 20 mm Marshall asphalt but 14 mm béton bitumineux pour chaussées aéronautiques (BBA) is selected for the surface course because of local preference.

The traffic is medium, so the preferred option would be 20 mm DBM, but 20 mm BBA is selected for the surface course because of local preference.

The preferred option for the base would be 31.5 mm BBA, which is selected for the base.

The choices of materials for this pavement are:

- Surface course – 14 mm BBA.
- Binder course – 20 mm BBA.
- Base – 31.5 mm BBA.

9.1.4 Cycleway Example

For cycleways that are not regularly overrun by other vehicles, go to Table 7.35 via Figure 7.9 for the surface course and Table 7.36 for the structural layer or layers.

The preferred option for surface course would be 10 mm SMA, but 10 mm DBM is selected for the surface course because of limited funds.

The preferred option for the structural layer or layers would be 20 mm DBM, which is selected for the structural layer(s).

The choices of materials for this pavement are:

* Surface course – 14 mm DBM.
* Structural layer or layers – 20 mm DBM.

9.1.5 Dock Example

From Figure 7.12 for storage areas of docks over strong foundations, go to Table 7.44 for the surface course, Table 7.22 for the binder course and Table 7.25 for the base.

The usage is high and there are no restrictions on funding, so the preferred option for the surface course would be 10 mm or 14 mm grouted macadam, of which 10 mm grouted macadam is selected for the surface course.

Similarly, the preferred option for binder course would be 20 mm EME2, which is selected for the binder course, and the preferred option for base would be 31.5 mm EME2, which is selected for the base.

The choices of materials for this pavement are:

* Surface course – 10 mm grouted macadam.
* Binder course – 20 mm EME2.
* Base – 31.5 mm EME2.

9.2 Summary

There is an impression that, currently, many nominal engineers select the mixture for any project based purely on the mixture used on the last project in order to avoid changing anything in the project documentation. Whilst this approach can work reasonably most of the time, it does not allow for either differences in the circumstances pertinent to the site or developments in the asphalt mixtures available. The methodology set out in this book shows a more complete and logical approach to examine the options available in each case, with the final decision being left to other less major aspects that are not completely covered here.

There is rarely a single asphalt mixture that is the only one suitable for a situation, but there are some mixtures that are more suitable than others. Similarly, many mixtures are suitable for many different situations but there are some situations in which they ae unsuitable. It is hoped that the review will help engineers and others who select asphalt mixtures, either as a specifier or producer, to consider the different possibilities available that may be more suitable than the one they regularly use without straying into using an unsuitable one.

Index

Figures are displayed in italics and tables are displayed in bold

Printed in the United States
by Baker & Taylor Publisher Services